Routledge Revivals

The Berlin Discussion of the Problem of Evolution

First published in 1909, this book collects the author's lectures on the 'problem of evolution' and the resultant debate. The first considers the validity of the Theory of Evolution and whether it is in opposition to the Christian view of creation. The second examines the assertion that evolution harmonises only with Monism rather than Theism and which of the two views is preferable. It also looks at the popular identification of Darwinism with evolution, if it is scientific and the results this leads to. The third looks at man's position in the problem of evolution — whether we are bound to bring in considerations higher than the zoological — and the evidence for our descent from 'brutes'.

Routledge Revivals

The Recent Discussion on the Problem of Evolution

[illegible]

The Berlin Discussion of the Problem of Evolution

Full Report of the Lectures Given in February 1907, and of the Evening Discussion

Erich Wasmann, S.J.

Routledge Revivals
Routledge
Taylor & Francis Group

First published in 1909
by Kegan Paul, Trench, Trübner & Co., Ltd

This edition first published in 2016 by Routledge
2 Park Square, Milton Park, Abingdon, Oxon, OX14 4RN
and by Routledge
711 Third Avenue, New York, NY 10017

Routledge is an imprint of the Taylor & Francis Group, an informa business

Publisher's Note
The publisher has gone to great lengths to ensure the quality of this reprint but points out that some imperfections in the original copies may be apparent.

Disclaimer
The publisher has made every effort to trace copyright holders and welcomes correspondence from those they have been unable to contact.

A Library of Congress record exists under LC control number: 10002339

ISBN 13: 978-1-138-65890-5 (hbk)
ISBN 13: 978-1-315-62048-0 (ebk)
ISBN 13: 978-1-138-65891-2 (pbk)

THE BERLIN DISCUSSION

OF THE

PROBLEM OF EVOLUTION

FULL REPORT OF THE LECTURES GIVEN IN FEBRUARY 1907, AND OF THE EVENING DISCUSSION

BY

ERICH WASMANN, S.J.

AUTHORISED TRANSLATION

KEGAN PAUL, TRENCH, TRÜBNER & CO., LTD.

DRYDEN HOUSE, 43 GERARD STREET, LONDON, W.

B. HERDER

17 SOUTH BROADWAY, ST. LOUIS, MO.

1909

The work entitled *The Berlin Discussion of the Problem of Evolution*, by Erich Wasmann, S.J., has been submitted to the censorship of members of the same Society, who have sanctioned its publication, and therefore we give permission for it to be printed, if it seems expedient to those whom it concerns.

In confirmation whereof this letter is signed with our hand, and sealed with our official seal.

ERNEST THILL, S.J.
Praep. Prov. Germ.

EXATEN, *July* 22, 1907.

PREFACE

An earnest desire has been expressed in various quarters that I should publish the lectures, which I delivered in Berlin in February 1907, and so render them accessible to those who did not actually hear them, and that I should append to them a report of the discussion to which they gave rise. I am the more ready to comply with this request as many persons who wished to hear the lectures in Berlin were unable to do so, all the tickets of admission having been sold a week before the delivery of the first lecture.

In the meantime a little work by Dr. Burdinski has appeared, entitled '*The Struggle in Berlin regarding the Cosmic Position of Man:* A detailed report of the lectures given by Father Wasmann, S.J., and of the evening discussion on them, with critical remarks.' Should, however, any one refer to this work for a full and impartial account of my lectures, he would be disappointed, and Dr. Burdinski's report of the evening discussion is still more unsatisfactory. This work is merely a superficial review, written with the partisan spirit which characterises the *Vossische Zeitung*,[1] nor did its publication

[1] In proof of this statement see the *Germania*, Nos. 79, 80, 81, of April 7th to 10th, 1907, where a criticism of Burdinski's work will be found. It does not call for further consideration here.

satisfy the requirements of many people, who wished to be supplied with trustworthy data from which to judge of my lectures and of the evening discussion. In fact, Dr. Burdinski's work increased rather than diminished the demand for another report.[1]

With regard to the circumstances that led to the delivery of my lectures in Berlin, I may state that in his lectures on the 'Dispute regarding the Theory of Evolution,' given at the Academy of Music in Berlin, in April 1905, Professor Haeckel of Jena referred repeatedly to my book entitled '*Modern Biology and the Theory of Evolution*,' in fact he stated that the appearance of this work had led him to deliver his lectures. It seemed therefore expedient, in view of the many misunderstandings to which Haeckel's references had given rise, to publish a definite statement of my own opinion. Such a statement had, it is true, been made in my 'Open Letter to Professor Haeckel,' which appeared in the *Germania* and in the *Kölnische Volkszeitung* on May 2nd, 1905, and the same Open Letter, with some additions, is printed in the appendix to the third edition of *Modern Biology and the Theory of*

[1] At the request of the editor of the *Umschau* (Frankfurt a. M.), in Nos. 14 and 15 of that review I gave a short sketch of the contents of my Berlin lectures. The editor has recently, in a number of German papers, referred to this sketch as being the author's first publication of the lectures. Whilst acknowledging the editor's courtesy, I feel obliged to take this opportunity of stating emphatically that the above-mentioned short sketch cannot be regarded as an 'authentic publication' of my lectures, especially as the sketch was not printed word for word as I wrote it, and some of the corrections which I made in the proof-sheets were not accepted.

Evolution, published in 1906. But the contents of Catholic papers and of scientific works written by Catholic authors do not, as a rule, penetrate to those who have been most influenced by Haeckel's lectures in Berlin, or if they reach them at all, it is in a distorted form. Therefore I deemed it very important to give a course of lectures in Berlin itself on the same subject.

The first suggestion that I should give such a course was made to me on July 21st, 1906, by the managers of the Sachs concerts, who had organised Haeckel's lectures in the previous year. I did not, however, fall in with the proposal, being anxious to avoid all appearance of provoking a discussion, and the invitation, which I eventually accepted, proceeded from another quarter, and was framed in such a way as to guarantee the scientific character of the lectures.

On January 25th, 1907, the syllabus of my lectures appeared. They were delivered in the upper Hall of the 'Philharmonie.' In this syllabus the following points were mentioned as topics for discussion:—

First Evening (February 13th).—The Doctrine of Evolution considered as a Scientific Hypothesis and Theory (with lantern slides).

Second Evening (February 14th).—Theistic and Atheistic Doctrines of Evolution—Evolution and Darwinism.

Third Evening (February 17th).—The Application to Man of the Theory of Evolution (with lantern slides).

On the evening of February 18th, in the great hall of the Zoological Gardens, there was a discussion on the subjects with which I had dealt in my lectures.

A fuller account of the circumstances which led to this discussion will be found in the preliminary remarks prefixed to the second part of this work.

The syllabus of these lectures was signed by the following gentlemen:—Dr. Horn, President of the German Entomological Society; Professor Dr. Kny, member of the Privy Council, and Professor at the School of Agriculture; Professor Kolbe, Curator of the Natural History Museum; Dr. Plate, Professor at the School of Agriculture; Mr. Rintelen (who is now dead), member of the Privy Council, and President of the Board of Higher Education (Oberlandeskulturgerichts); Dr. Waldeyer, member of the Medical Council and permanent secretary to the Academy of Science.

The cards of admission to the lectures were on sale at Herder's Library in Berlin, at the price of a shilling for each evening—reserved seats, two shillings. Students could obtain tickets for sixpence on application to the porter of the School of Agriculture and at the Reading Room of the Academy. It was falsely reported in some of the

Berlin newspapers, and especially in the *Vossische Zeitung* of February 12th, that 'two-thirds of all the seats were assigned in advance to Catholic associations and students' societies, which were to start the applause.'

The inaccuracy of this statement becomes apparent, when we consider that there are scarcely three hundred members of all the Catholic associations and students' societies collectively, whereas there were at least one thousand persons present at each of the lectures, and about two thousand at the evening discussion. Moreover, this false report published by the *Vossische Zeitung* was immediately contradicted by Professor Plate,[1] a member of the Committee, but his correction was refused by the editor of the paper in question.

At the request of the organising Committee, before beginning my second lecture, I drew attention to the falsehood of the assertion that any partiality had been shown in the distribution of the admission tickets, but, nevertheless, the report held its ground, and was referred to again by Burdinski in his *Struggle in Berlin regarding the Cosmic Position of Man* (p. 2), which deals with my lectures.

There is another point to which I must allude. The syllabus of my Berlin lectures had been drawn up long before the dissolution of the German Parliament. At the end of November, 1906, the great

[1] As Professor Plate is a member of the German Monistenbund (Monistic Society), it is impossible to accuse him of being one of my partisans.

Philharmonie Hall had been engaged by my agents in Berlin for the days mentioned in the syllabus, whereas the dissolution took place on December 13th. It is therefore quite impossible that the Jesuits should have had any previous knowledge of the new political combination, and yet Dr. David v. Hansemann, who was formerly Rudolf Virchow's assistant, contributed to the *Vossische Zeitung* of February 26th, 1907, an article signed with his name, and entitled 'Quidquid id est, timeo Danaos et dona ferentes!' From it I quote the following passage:—

'Herr Wasmann is, in fact, the branch smeared with bird-lime which is designed to catch a multitude of bullfinches, who set the tone of public opinion. People are to be induced to believe that men who have such liberal opinions cannot possibly constitute a public danger. I will even go so far as to express the opinion that the Jesuit Order did not send Herr Wasmann here with this general intention, but that the actual political situation suggested to the Jesuits the idea of despatching him to Berlin at this particular time. Apparently he brings with him a gift, a form of scientific enlightenment, a gift intended to serve as an intellectual bribe, giving men an impulse in a direction to which hitherto most energetic resistance has been offered. By means of this wooden horse, the Jesuits hope to gain a footing in the land in which, above all others, the freedom of science has been most highly respected. We shall fare as did the Trojans of old. If this gift brings our enemies into the country, they will do their utmost to take root and to exert their injurious influence with the reckless tendency to destruction which characterises them, just as did the Greeks in Troy. Therefore again I say: *Quidquid id est, timeo Danaos et dona ferentes!*'

One may well feel inclined to ask: 'What must educated men think of a Berlin Professor who seriously gives utterance to such fantastic ideas?' Can Dr. v. Hansemann fail to see that his words contain a serious insult to his colleagues at the University of Berlin, who signed the invitation to my lectures, and so apparently helped to ensnare finches? We receive, moreover, a strange impression of scientific toleration, when von Hansemann demands not only that people, whose views differ from his own, should be forbidden to speak, but that they should actually be excluded from the country.

The present work is divided into two parts. The first part contains the three lectures given in Berlin.

The second part contains a report of the evening discussion, with critical remarks.

The lectures in Part I. were taken down in shorthand with the greatest possible accuracy, and these shorthand notes, in somewhat abbreviated form, appeared in the *Germania*, Nos. 38, 39, 41. A long passage in the first lecture, which was illustrated by fifty-six lantern slides, is necessarily much condensed, as the illustrations cannot be reproduced here. Any one who takes an interest in the facts represented can find detailed information and diagrams relating to them in the tenth chapter of the 3rd edition of my book entitled *Modern Biology and the Theory of Evolution*, Freiburg i. Br., 1906.[1]

[1] I shall refer to this work henceforth as *The Modern Biology*.

While the first lecture appears therefore in a considerably abbreviated form, at the beginning of the second and third lectures introductory matter has been inserted, to facilitate the comprehension of what follows.

In Part II. a short historical introduction has been prefixed to the account of the proceedings at the evening discussion. Then come the speeches of my opponents, and, as far as they dealt with the subject-matter of my lectures, they are accurately reported and critically examined. My answer stands as it was taken down by the shorthand writer, with the addition of a few comments. Then follows a short supplement, summing up the results of the discussion on the lines of the newspaper reports.

I fear that many who read these Berlin lectures will be disappointed at the plain, dry style in which they are worded, and will ask why Fr. Wasmann did not speak with the same kind of 'inspiring eloquence' which Haeckel employed in his lectures. The answer is simply that my words were not addressed to the imaginations and emotions of my audience, but to the clear, cool judgment of the intellectual men of Berlin assembled before me. I aimed at throwing real light upon the important question: 'What are we to think of the doctrine of evolution?'

The consideration of this great question falls naturally into subdivisions, dealing with single points, which were discussed in the three lectures.

Lecture I.—What claim for consideration has the doctrine of development (or evolution) regarded as a scientific hypothesis and theory? How far is it firmly based on facts? Is it, or is it not, in opposition to the Christian view of creation?

Lecture II.—Is the Monistic assertion true, that a scientific doctrine of evolution harmonises only with Monism, and not with Theism, as regards the view taken of creation? Which of the two views is preferable in the case of a scientist who is capable also of philosophic thought? What account can be given of the popular identification of Darwinism with the Theory of Evolution? Is it scientific or not, and to what results does it lead?

Lecture III.—What is the position of man in the problem of evolution? Is it permissible to consider this question from a purely zoological point of view, or are we bound to bring other and higher considerations to bear upon it? What are the zoological and palæontological proofs of the descent of man from brutes?

The object of devoting an evening to discussion, after the delivery of the lectures, was to elicit a practical statement of the scientific divergencies in opinion between myself and my opponents, for

I hoped in that way to secure some approximation of our respective views. How far this end was attained, the reader will be in a position to judge, from the second part of this work.

In the Supplementary Note will be found my remarks upon Professor Plate's reply, which has only just been published, and could not therefore be noticed in the Supplement.

CONTENTS

PAGE

PREFACE v

PART I

LECTURES ON THE PROBLEM OF EVOLUTION

I. THE HYPOTHESIS AND THEORY OF EVOLUTION IN NATURAL SCIENCE 1

II. THEISTIC AND ATHEISTIC EVOLUTION—EVOLUTION AND DARWINISM 21

III. THE DESCENT OF MAN 49

PART II

EVENING DISCUSSION ·

Introductory Remarks on the Origin of this Discussion . 85

Opening Speech by Professor Waldeyer . . . 90

Speech of the First Opponent, Professor Plate . . 92

„ Second „ Dr. Bölsche . . . 117

„ Third „ Professor Dahl . . 123

„ Fourth „ Dr. Friedenthal . . 134

„ Fifth „ Professor von Hansemann . 144

„ Sixth „ Count von Hoensbroech . 158

„ Seventh „ Mr. Itelson (the Author) . 159

„ Eighth „ Dr. Juliusburger . . 160

„ Ninth „ Dr. Plötz . . . 176

„ Tenth „ Dr. Schmidt-Jena . . 187

„ Eleventh „ Dr. Thesing . . . 201

Speech of Father Wasmann (close of the Discussion) . 209

SUPPLEMENT 240

SUPPLEMENTARY NOTE upon pp. 171, 218, 231 . . 258

INDEX 261

LECTURE I

(*February* 13, 1907)

THE HYPOTHESIS AND THEORY OF EVOLUTION IN NATURAL SCIENCE

LADIES AND GENTLEMEN,—

There is certainly much truth in the statement, made in the syllabus to these lectures, that a keen interest is taken in Berlin in the problem of evolution. The mere sight of my distinguished audience is sufficient confirmation of it, and there is abundant justification for this interest, as the question: 'What are we to think of the theory of evolution?' is one that attracts universal attention. All are agreed in their interest on this subject, but the answers given to the question are very various, in fact, in many respects actually contradictory. On one side we are told that the theory of evolution is a mere hypothesis devoid of real scientific support, that it is based on no definite facts, and in its further development is an outcome of atheism, which aims at the overthrow of Christianity. On the other side we hear that it is most firmly and satisfactorily supported by scientific facts, and is already so far developed as to claim the recognition of every biologist, let his views on religion be what they may. Finally, we are told by a third party that the theory of evolution is in direct antagonism to Christianity, and is, as Haeckel says, 'the chief weapon employed against Christianity by the heavy artillery of Monism.'

Which of the three statements is true? I am inclined to believe that all are to some extent both true and false. Our first business must be to formulate our ideas clearly. *Clearness always leads the way to truth,* therefore whoever seeks truth must first aim at clearness. On this account I regard it as the object of these lectures to explain, as clearly as possible, the various ideas which we have to distinguish in our consideration of the theory of evolution, and to show in what light it behoves us to regard these ideas, and what ought to be our attitude towards them.

Do not imagine that I have come here to argue against Haeckel's views. Such is by no means my intention. In July of last year the Sachs Concert Committee invited me to come to Berlin in order to read a paper against Haeckel. I declined the invitation, as it did not seem to me expedient to appear in direct personal opposition to him, and I had no wish to increase the already existing excitement. More than enough antagonistic speeches have been delivered already, and I only wish to throw some real light on the subject, trusting in this way to do a good work.

The general plan of my lectures has been sketched in the programme. In the first I hope to give a brief account of the doctrine of evolution under the aspect of a scientific hypothesis, and to illustrate my remarks by a series of photographic lantern slides, which I have selected from my special province, viz. the guests or inquilines (*i.e.* parasites) of the ants and termites, or white ants, because with regard to these creatures I feel most at home, and have not to rely upon the authority of others, but incidentally I shall refer to arguments derived from other departments of science.

In my second lecture I intend to distinguish, as clearly as I can, between the doctrine of evolution, as a scientific hypothesis, and the same doctrine as a philosophical theory of life, and further, between the doctrine of evolution as resting upon a theistic foundation, and as based upon materialistic atheism, and finally I wish to distinguish between Darwinism and the theory of evolution.

In my third lecture I intend to deal with the application to man of the theory of descent, and to show you a few diagrams by way of illustration. We have arranged to have an evening discussion on Monday in the great hall of the Zoological Gardens.

When the struggle arose between the Copernican and Ptolemaic systems, over 350 years ago, people had no conception how far the new ideas would lead. It was reserved for more recent ages to connect the heliocentric conception of the universe with the natural evolution of our solar system and the uniform development of the Cosmos as a whole, including all the heavenly bodies.

In this vast universal evolution, which we assume to have taken place, and which is based on natural laws, the evolution of our little world occupies a scarcely perceptible period of time, barely a minute, and of this minute a small fraction, (that nevertheless, according to geologists, lasted millions of years), was occupied by the evolution of organic life before the appearance of man.

The progress made in zoology, botany, and especially in palæontology has led scientists to investigate more closely the relation existing between our present animals and plants and the extinct fossil varieties. Are we to regard the animal and vegetable kingdoms, as we know them, as something fixed and unchangeable, or are our present animals and plants modified descendants of older, and for the most part extinct, ancestors, which have been to some extent preserved to us as fossils ?

Two kinds of answers have been given to these questions. The advocates of the theory of permanence maintain that the systematic species recognised at the present day in our zoological and botanical systems remain unchanged.[1]

[1] That the theory of permanence in its historical form assumes the unchanging nature of the systematic genera is well known to every student

The facts furnish us with no evidence proving that modifications of the species extend beyond the limits of those genera, and we are not justified in speaking of any evolution of species from one another, nor of any stock as common to the species now existing or to those that existed formerly.

The theory of evolution, on the contrary, leads its supporters to declare that the fauna and flora of the present day ought to be regarded *as the final outcome of a previous evolution*, and to some extent as the final function of a long differential and integral calculus of nature. Thus the question may be worded as follows:—*Has an evolution of the organic species from the original stock taken place, or not?*

It will be easily seen that we have no right to say that this theory of evolution is the product of Atheism. The question whether there is a probable historical connection between the present and the fossil forms of animals and plants, is a purely scientific one, arising logically from the researches made by zoologists, botanists, and palæontologists. I wish therefore to lay emphasis

of modern zoology and botany (see *Modern Biology*, pp. 261, 303, 315, etc.). As examples of systematic species I may mention the lion, the tiger, and the jaguar within the genus *felis*. The characteristics distinguishing the species are essential only in the empiric and not in the philosophical sense. For this reason some scientists subsequently introduced the idea of natural species, comprising groups, varying in size, of systematic species. *This* theory of permanence, if compared with the historical theory of the permanence of the systematic species, is, however, already a restricted theory of evolution (cf. *ibid.* p. 294).

on the statement, that the doctrine of evolution as a *scientific hypothesis and theory* naturally arose from our increased knowledge of zoology, botany, and palæontology.

What is then the subject-matter of the doctrine of evolution as a scientific hypothesis and theory? As I have already suggested, it is primarily to investigate the succession of the forms of plants and animals, since the first appearance of life on our globe, in order to classify species, genera, and families; and in the second place *to explain this order by a natural evolution of species.* The object therefore of the science of natural evolution is *the investigation, both as to facts and to causes, of the lives of successive organic forms, which terminate in the now existing species.*

And what is *not* the object of the doctrine of evolution? It is *not* its object to explain the origin of life upon this earth. The question whether we must assume spontaneous generation or creation, in order to account for the coming into existence of the first organisms, is a philosophical problem, outside the scope of the scientific theory of evolution, and not belonging to it. In my next lecture I intend to discuss these metaphysical problems—for to-day I must limit myself to the doctrine of evolution as a scientific hypothesis and theory.

This doctrine of evolution is obviously *not an experimental science*; it is a hypothetical construction which reduces itself to *a theory.* It is only capable of giving us a higher or lower degree of *probability* as to the processes of historical development, but the evolution of a race does not admit of being demonstrated by observation or experiment. Nothing else indeed is possible, for man appeared upon the earth as an epigonos (*i.e.* an after-born), who came into existence at the close of a course of evolution that had lasted millions of years. When he looks back he sees only monuments, ruins, and traces of previous evolutions; and he cannot even survey the evolution itself, he can only obtain some conception of it by way of inference, after a careful and impartial comparison of very various pieces of evidence.

The doctrine of evolution is therefore *not an experimental science*, and can never be one. It is essentially *a theory based upon a group of hypotheses* which are in harmony with one another, and afford *the most probable* explanation of the origin of organic species. We cannot demand to see the evolution of species taking place before our eyes, in such a way as to give us a direct confirmation of the theory of evolution. Man was born far too late, and lives far too short a time, to be able to make such a demand. Imagine a fly, destined to live but one day, which comes to life one beautiful morning in spring, and sees all around it the trees in full blossom. That the blossoms came forth from buds which gradually unfolded, and that the blossoms in their turn will lose their petals and develop into fruit, all this must remain hidden from the fly during its few hours of life. It might therefore be tempted to believe that the blossoms all around were created by the good God exactly as it sees them, and will remain unchanged for ever. The fly would be greatly mistaken, and even as an ephemeral fly, if it had intelligence, it might perceive some slight signs that the splendour of the blossoms was not unchanging. It might see that, in the course of a few hours, some buds had already opened more fully, some blossoms had lost their petals either partially or wholly. The opening buds are those rare traces of modification of species which we can still prove to have taken place, although within comparatively narrow limits. If we continue the simile, the falling petals are the species in process of extinction, and the fallen leaves are the extinct species known to us only as fossils, which reveal to us the fate of all organic species on earth;—they come and go and give place to their successors, and though the duration of their existence may be reckoned in thousands or even millions of years—as is that of many kinds of the Brachiopod genus *Lingula*—yet for them, as for each one of us, there is a beginning and an end. But let us now abandon the simile.

Upon what evidence is the doctrine of evolution as a scientific hypothesis and theory based?

We must distinguish two kinds of evidence,

direct and indirect. The direct proofs are those faint traces of transformation of species, as they may still be discovered; such, for instance, as the botanist Hugo de Vries has described in support of his theory of mutation. He shows that in the botanical genus Œnothera, mullein, new forms are still being developed, which behave like real species. Against this theory of mutation it has been urged with some reason that the mutation is less important than de Vries believed. Standfuss has established, by means of numerous experiments in breeding butterflies, that mutation has scarcely any significance as a factor in the formation of species. Standfuss is of opinion that it is scarcely possible for species to be formed by fluctuating variation (which includes Darwin's accidental modifications). He regards as of real importance only the adaptive variations, *i.e.* modifications due to accommodation, which are caused by definite external causes and are transmitted to succeeding generations. What we mean by modifications, due to accommodation, and how they can produce new species and genera, I hope to show you later on by means of photographs from my own special department.

I need only say a few words about the indirect proofs. With regard to them the scientist, who wishes to arrive at a conclusion on evolution of species, behaves like a skilful public prosecutor, who wishes to secure the conviction of a prisoner charged with committing an offence which was witnessed by no one. The prosecutor collects *circumstantial evidence* in all directions against the accused man, and the greater its amount, the more closely can he press home the accusation. We must deal with the indirect evidence adduced in support of the theory of evolution in the same way. It is derived from comparative morphology, comparative anatomy, comparative history of the evolution of the individual, comparative bionomy, geography of animals, and especially from palæontology. I will refer at once to a few instances derived from this last source of evidence. There are hundreds of kinds of ants, which we know through their having been preserved to us in the tertiary amber of the Baltic and Sicily. Amongst them occur several genera which still exist, but scarcely a species that is identical with the present ones. We can hardly avoid coming to the conclusion that our ants are the *descendants* of these fossil varieties, and that they have come into being by way of natural evolution of the race, and not by way of a new creation.

Again, if we compare the fossil termites of the tertiary epoch with those now known to us, we are forced to assume that the latter are modified descendants of the former, and that they have come into being by way of natural race evolution, not by way of *a new creation*. Further, if we consider the oldest of the still existing varieties of termites, viz. the Australian genus Mastotermes, and compare the formation of the wings with that of the Blattidæ, or cockroaches, both fossil and still existent, we shall probably find that the termites in some prehistoric palæozoic age were evolved from one and the same stock as the ancestors of our present black-beetles.

I might give many such instances, but it is time for me to pass on to my photographs. They represent the inquilines, living among ants and termites, and we shall observe a number of interesting phenomena, which are biologically explicable only from the point of view of evolution.

[At this point some photographic lantern slides were displayed.[1]]

To illustrate the statement regarding direct proofs the lecturer displayed photographs of the varieties of the Dinarda, a kind of beetle which lives with the ants, and is still producing new forms. He also showed species of the genus Doryloxenus, which in comparatively recent times in the East Indies have ceased to live with the ants and have become guests of the Termites, thus being changed into new systematic varieties. The same change was shown to have taken place in some African species of the genus Pygostenus.

Subsequent photographs illustrated the indirect means of justifying the theory that new species, genera, and families are formed by variation due to accommodation. These photographs also were chosen from the department of science which the lecturer has made especially his own. Most of them represented brachyoptera, beetles with very short

[1] For further details and illustrations, see *Modern Biology*, chap. x. pp. 323-431, and also 'Instances of Recent Formation of Species among the Inquilines of Ants and Termites,' *Biol. Zentralblatt*, 1906, Nos. 17 and 18.

wing-cases, which have been admitted as guests by the wandering ants and the slave-keeping ants. The guests of these ants can be divided into three biological classes: (1) the true guests, the most interesting type of which is an African species named *Sympolemon anommatis.* As an example of the second class, a photograph was shown of a brachyopteron known as the *Mimeciton pulex,* or ant-ape, which lives among the blind wandering ants of Brazil, and is enabled to pursue its parasitic existence through bearing a remarkable resemblance to the ants in the form of its antennæ and of its whole body. The ants' sense of touch in their antennae is deceived by this resemblance. In connection with this subject of mimicry, other illustrations were given. As representing the third biological class, the lecturer displayed a photograph of the so-called hostile type (Trilobitideus, etc.) in which the beetle is protected by a shell against any attack on the part of ants. These creatures have been evolved out of the same family of Brachyoptera, although in form they differ from one another more than an ape differs from a tortoise. The same final result might be produced by very various methods of adaptation; so in some cases synechthry has led to the development of a genuine guest relation, which in other cases is due to mimicry.

A series of photographs illustrated the highly interesting evolution of Pselaphidæ to Clavigeridæ. All true inquilines have organs of exudation, developed in a greater or less degree, which emit an aromatic substance that the ants enjoy licking, and it is in order to obtain it that they feed their guests. The formation of these organs of exudation is particularly interesting in the case of certain Paussidæ, which are allied to the Carabidæ, or ground-beetles. Their antennæ have been transformed into veritable cups containing a sweet fluid. The theory as to the origin of slavery among ants was explained by another series of photographs, at the conclusion of which was a picture of a peculiar parasitic ant (*Anergates atratulus*), which possesses no real workers, but only winged females and wingless, strangely degenerate males, so that this ant is in absolute dependence upon the workers of the wood ants, amongst whom it lives. This parasitic ant must be descended from a genus that possessed workers and once led an independent existence.

As an instance of discontinuity of variation, the lecturer showed a photograph of the males of the Formicoxenus, that closely resemble workers, and they are the more remarkable because no trace can be found among ants of any gradual transition from the male to the female worker.

The last series of photographs showed a number of strange guests among the termites, belonging to various families of beetles and diptera. By accommodation to the way of life among the termites, new systematic families have arisen (*Termitoxeniidæ* and *Thaumatoxeniidæ*) in exactly the same way as they arose among the beetles (*e.g. Clavigeridæ, Paussidæ,* etc.) in consequence of accommodation to the myrmecophile life. The last photograph showed a coloured longitudinal section of the *Termitoxenia Assmuthi,* a very small fly living with the white ants of the East Indies.

The lecturer closed his display of photographs with approximately the following words:—

I wish to draw your attention to the fact that accommodation to the life of ants and white ants or termites has in all probability led to the formation of *new species, genera, and families* among their guests, which belong to very various families and orders of insects. In some cases (*Thaumatoxena*) the characteristic marks have been so completely altered by accommodation that it is scarcely possible for us to determine to which order of insects this strange creature belongs. In other cases (*Termitomyia*) the whole development of the individual is modified in such a way that it resembles that of a viviparous mammal rather than that of a fly. The oft-repeated assertion of the upholders of the theory of permanence, that variation by way of accommodation only produces abnormal forms within the species, is thus seen to be false.

What conclusions are we to draw from these considerations? If we carefully study the phenomena, which have just been presented to us, we must acknowledge that only the theory of evolution can explain to us how these interesting forms came into being. We cannot supply a scientific explanation by merely declaring that these strange little creatures, such, for instance, as the *Mimeciton*, or ant-ape, were created by God expressly for this or that variety of ant. *The principle of the theory of evolution is the only one which supplies us with a natural explanation of these phenomena*, and therefore we accept it. But to what extent are we to accept it? *Just as far as its application is supported by actual proofs.*

If I were to attempt to answer the question how far this is the case, I should have to refer to many other examples from other branches of science, but I think it is possible to establish the following statement, as the result, not merely of my own investigations, but of those of others, who, like myself, have devoted close attention to the phenomena of accommodation in general, and to phylogenetic evolution in particular.

In the case of the species of the same genus, the genera of the same family, and often for the families of the same order,—even for orders of the same class, *the probability is in support of evolution*, and we meet with actual points of contact proving the relationship between the various forms. But the higher we ascend in the systematic categories, and the more closely we approach the great chief types of the animal world, the scantier becomes the evidence; in fact, it fails so completely that we are finally forced to acknowledge, that *the assumption of a monophyletic evolution of the whole kingdom of organic life is a delightful dream without any scientific support.* The same may be said of the assumed monophyletic evolution of the whole animal kingdom on the one hand, and of the whole vegetable kingdom on the other, from one primary form respectively.

We have no *scientific* evidence to support these assumptions, such as we have for the relationship of species, genera and families, and we cannot prove anything without evidence.

Some one may charge me with having simply given utterance to Fleischmann's views. No, I do not follow Fleischmann in this respect, for he goes too far in his opposition to the theory of evolution, but he is right in saying that *it is impossible to trace back the chief types of the animal kingdom to one primitive form.* All attempts in this direction have failed. Fleischmann is not alone in making this assertion, for it is put forth by many other and far more important authorities. I should like to mention particularly Professor Oskar Hertwig, who, in the last chapter of his excellent handbook of comparative and experimental evolution, has discussed, in a very clear and logical manner, the evidence in favour of the theory of descent, which has been hitherto adduced from comparative morphology and evolution. He says: '*Evidence of the monophyletic development of different races is altogether wanting*, and we are forced more and more to accept the theory of development from a variety of stocks.'

Professor Boveri, who certainly was also quite free from 'theological bias,' in his last presidential address at the university of Würzburg, dealt with the history of organisms. He too regards it as impossible to trace back all the varieties of animals to one primitive form.

Von Wettstein among the botanists, and, more particularly, Steinmann, Koken, and Diener among the palæontologists, have recently come forward as champions of the theory of polyphyletic evolution, consequently no one can charge me with upholding it in my capacity as a '*theologian*.'[1] I abide by my conclusions, with just as much justification from the zoological point of view, as do the above-named eminent scientists, who, without being theologians

[1] See Diener's 'Palæontology and the Doctrine of Evolution' in the Austrian *Rundschau*, xi., 1907, No. 3.

or Jesuits, have declared in favour of polyphyletic evolution.

Allow me for a moment to view the subject from the aspect of Christianity and to refer to the biblical account of the creation. There we read (Gen. i. 11-25) that God created beasts and plants, each according to its kind. This biblical expression is not to be judged by the standard of modern zoology. That the geological development of the world is not irreconcilable with the biblical account is universally acknowledged by theologians, and it would seem that the same opinion is gaining weight as to the development of the organic world.[1]

We must first of all state clearly that *the Bible is not intended to instruct us in modern science*, and we scientists of the twentieth century ought not to seek zoological information in it. The Bible is meant to give instruction, not on science, but on the way of salvation, as Leo XIII. proclaimed in his beautiful encyclical *Providentissimus Deus.*[2]

The biblical account was compiled for the information of men of every age, independently of the changing theories of human science.

When we read the stately account in broad outlines which Holy Scripture gives of the Creation, and when we are told that God's creative word produced the various kinds of plants and beasts

[1] Cf. P. Knabenbauer in *Stimmen aus Maria-Laach*, xiii., 1877; *Glaube und Deszendenztheorie* (Faith and the Theory of Descent); also *Modern Biology*, p. 255, etc.

[2] Cf. the quotation in *Modern Biology*, p 446, note 1.

from the water and the earth, we must not imagine that we have here any scientific definition of the idea of species. This idea, which Aristotle calls the εἶδος, was the offspring, many centuries later, of a definite system of philosophy, and much later still arose the scientific idea of species, which is so closely interwoven with the theory of permanence, and which was evolved by Ray, Linnæus and Cuvier in the eighteenth and nineteenth centuries.

If, therefore, modern science shows us that we probably ought to group the systematic varieties of the present and of the past together, so as to form genealogical pedigrees, we, as philosophers, may describe such pedigrees as 'natural species.' But we must not read this idea of species into the biblical account of the creation, as if it really occurred there. We can only say that, if this idea of kind is confirmed, it will be additional testimony to prove that the biblical account of the creation does not contradict the facts ascertained by science.

Personally I am firmly convinced that *the doctrine of evolution, considered as a scientific hypothesis and theory, is not at variance with the Christian theory of life*, although the contrary is often asserted.

We have just mentioned the natural kinds, which are identical with the sequences in evolution or the pedigrees of the theory of descent.[1]

[1] See further *Modern Biology*, p. 303 *et seq.* The misconceptions there refuted reappeared in the course of the evening discussion, in the speech made by Professor Plate, my chief opponent.

No expert is yet able to say how many such genealogical series we must assume; perhaps we shall learn more about it two thousand years hence. Nor do we know anything about the hypothetical primitive forms, which are the starting-points of the genealogical series. We are still quite in the dark too as to the laws governing their evolution. All these are biological problems which must be investigated in the centuries to come.

But I must now draw to a close. If we assume that God is the creator of all things, and that the world created by Him has evolved independently and automatically, we have actually a greater idea of God than if we regard Him as constantly interfering with the working of the laws of nature. Let us imagine two billiard-players, each having a hundred balls to direct. The one needs a hundred strokes in order to accomplish his end, the other with one stroke sets all the balls in motion, as he will. The latter is undoubtedly the more skilful player. St. Thomas Aquinas stated long ago that the force of any cause was the greater, the further its action extended. God does not interfere directly in the natural order where He can work through natural causes. This is by no means a new principle, but a very old one, and it shows us that the theory of evolution, as a scientific hypothesis and theory, as far as it can be really proved, is perfectly compatible with the Christian theory of the origin of things. According to this view, the evolution of the organic world is but a little line in the millions of pages contained in the Book of the Evolution of the whole universe, on the title-page of which still stands written in indelible letters: '*In the beginning God created Heaven and Earth.*'

LECTURE II

(*February* 14, 1907)

THEISTIC AND ATHEISTIC EVOLUTION—EVOLUTION AND DARWINISM

Ladies and Gentlemen,—

I must ask the favour of your closest attention this evening, because my present subject is both difficult and dry. I do not intend to describe to you in glowing colours the theistic theory of life with its exalted idea of God, which forms the highest and noblest object of human knowledge, nor do I propose to appeal to your emotions and to impress upon you the advantages possessed by Christianity, in virtue of the doctrine of immortality, over the depressing monistic theory, according to which death ends everything for us, and we continue to live only in the atoms which once made up our mortal bodies. I wish rather to appeal to your intellect, and to try to lay clearly before you the most important conceptions, which are often confused with one another by those who discuss the doctrine of evolution. For this purpose I must refer first to the philosophical doctrine of evolution, in order to bring out clearly the points of contrast in the various opinions on this subject. We shall then proceed to a critical comparison of the doctrine of evolution and Darwinism.

In our last lecture we considered the doctrine of evolution as *a scientific hypothesis and theory,* and we found that it was applicable to *the investigation into the actual condition and origin of the genealogical series of living organisms,* which reach from the most remote palæozoic age to the present time, when we find the living varieties of our own day to be the latest offshoots of these stocks.

It follows directly that this theory of evolution has nothing to do with the theory of the universe as such. One may be a monist or one may be a theist, and still equally well support this doctrine of evolution in the domain of natural science.

A further consequence is that it is incorrect, and not in accordance with the actual facts, to use the doctrine of evolution as a kind of battering-ram against Christianity. The monists do this, and, for the last forty years, Haeckel asserts that he has been using this theory, as 'heavy monistic artillery' against Christianity. *The scientific theory of evolution is in itself absolutely unconcerned with any theory of the universe,* although the human intellect feels a craving to bring this particular theory into touch with one or other of the various general views of the universe.

Theistic and Atheistic Doctrines of Evolution.

As soon as an intellectual man feels attracted by any scientific theory, his impulse is to give it a general application. Even if his actual knowledge reaches only to a definite point, and although in all probability much will remain unfathomed by scientific research even in the future, he is nevertheless inclined to entertain bright hopes of success, and mentally to expand his vast conceptions by means of arguments based only on analogy. Thus a *philosophical* theory of evolution arises out of the *physical* theory, and no legitimate objection can be raised to it in itself, for it only corresponds to the requirements of the human intellect. We were, in fact, touching upon philosophy in the first lecture, when I asserted that the Christian theory of the universe was not incompatible with the scientific theory of evolution. In making this assertion I at once imported a philosophical element into the subject, and in order to sketch a bold outline of the evolution of the universe according to natural laws, I generalised from a form of evolution proceeding from the first word of creation. Therefore it is quite according to nature for us to generalise about the physical theory of evolution on philosophical lines; but if we do this, we soon begin to ask on what foundation this generalisation can be constructed, and here we reach the domain *of the theories of the universe*, or, more accurately, of the various opinions regarding the origin, nature, and final object of the world (Weltanschauung).

It is impossible for such a theory to be independent of all presupposition. The theistic view is often blamed for not being 'free from presuppositions,' inasmuch as it assumes the existence of a personal Creator, and also the act of creation—but no Weltanschauung is free from hypotheses, not even the monistic, which assumes the indestructibility of matter and many other things, which cannot be proved, and which are no more connected with actual facts than are the presuppositions of the theistic theory.

In speaking of monism we must be careful to avoid confusion. There is a *scientific monism*, better called causalism, which seeks natural causes for every natural phenomenon, and requires these causes to be as simple as possible.

In this sense I too am a monist. As far as the natural order goes, I too desire to have a natural and, if possible, a simple explanation of every natural phenomenon, and we need not now take into account the supernatural order, as it does not fall within the scope of these lectures. Therefore in the sense which I have defined, I have no quarrel with monism. But the word has another signification: there is a *metaphysical monism, which asserts the essential identity of God and the world*, and in this sense Monism is used in contradistinction to Dualism.

Dualism, with which Christianity is charged, maintains that God is essentially distinct from the world. Monism declares that its God is essentially identical with the world. We need not ask how much is left for 'God,' when we have subtracted the 'world' from Him, as if we were working a sum in algebra—obviously the remainder—the actual being of God—is nothing. In this way monism turns out to be mere atheism, when we examine it closely.[1]

The kind of monism with which we are concerned this evening is therefore *metaphysical monism.*

Our next business is to state shortly the postulates and axioms of the two theories, to compare them with one another, and to test their relative values.

First let us take the postulates and presuppositions of the *monistic theory.*

1. We must not accept the existence of any personal Creator, of any so-called extramundane God, but assume the existence of the world with its laws from all eternity.

2. Dependent upon this is the further postulate: In order to account for the origin of the first organisms, we must assume a 'spontaneous generation,' *i.e.* a spontaneous development of the first organisms from inorganic matter.

3. We must not assume any kind of conscious

[1] Cf. on this subject my remarks on Dr. Schmidt-Jena's speech in Part II. The distinctions between Theism, Deism, and Pantheism are there brought out more clearly.

purpose or tendency on the part of living organisms. All has developed in strict conformity with mechanical laws.

4. We must not admit any essential difference between men and brutes. There can be no suggestion of the existence of a spiritual, immortal soul in man.

In opposition to these postulates, the *theistic view of creation* asserts the following :—

1. We are obliged to start with assuming the existence of a personal Creator, a Being who, in virtue of His infinite perfection, exists and has existed for all eternity through and of Himself; who, in virtue of His infinite perfection, contains in Himself the reason of His existence, whereas matter cannot contain in itself the reason of its existence. This God, as conceived of by Christianity, is present in all creatures; He is not far from us; He is, moreover active by His co-operation in all creatures, not as *deus ex machina,* but participating in the actions of all creatures, through His interior presence.

Perhaps the supporters of monism will declare these to be purely monistic ideas—but such is not the case. Monism has borrowed from theism these ideas of the universal presence of God, and of His co-operation in all activity on the part of creatures, and it has decked out its conception of God with them. The borrowing is on the side of monism. I need not dwell upon the distortion into a caricature of the theistic idea of God, which has been the work—I will not say of all the representatives of monism—but of many of them, and especially of Haeckel. We might almost say that Haeckel derived his conception of personality from the siphonophores or other cormi (animals fixed on one stock) of which the part-individuals are specially adapted to perform particular tasks; some serve to devour food, others swim, others afford protection, etc. Haeckel describes these individuals forming parts of one whole as swimmers, feeders, etc. This zoological conception of personality is certainly not applicable to God, for God cannot be a finite, corporeal being.[1]

Haeckel had not far to go to find his conclusion, which is,—there can be no personal God, for if there were one, he would be 'a gaseous vertebrate'! This is indeed philosophically untenable. Christian philosophy and theology have always formed a totally different conception of the personality of God. *God is the absolutely perfect entity,* absolute intelligence and absolute perfection with all its properties and without any separation of these properties from one another. *The personal God is fulness of being existing in itself and of itself.* And just because God is fulness of being, He can by His will (which is not, however, anything apart

[1] However, Haeckel and his followers are well known to have applied the zoological conception to God, as did also Dr. Plötz, in the course of the evening discussion, when he referred to God as an organism, and then asked for a 'Creator of the Creator.' Cf. my remarks on the speeches made by Dr. Plötz and Mr. Thesing (Part II.).

from His being) *call into existence all finite being* which, through its accidental and limited character, could not have in itself the reason of its own existence. This is the true theistic conception of God.

2. The theistic theory of the world involves the idea of creation. I grant that this presents a difficulty to our imagination. We cannot picture to ourselves by means of our fancy how anything, that previously did not exist, can come into existence and be produced out of nothing. It would be impossible unless an infinitely perfect Being existed, *virtually* containing beforehand the finite being in Himself.

This harmonises with the theistic conception of God, and so the idea of creation offers no internal philosophical contradiction.

3. Further, the theistic view, taken in conjunction with that of the creation of matter, lays down as its foundation the subjection to law of the whole cosmic evolution and of the entire evolution of the inorganic world, asserting that the first combination of atoms or electrons contained the definite material disposition from which, in the course of the succeeding millions of years, all the various constellations of atoms were to result by way of natural evolution.[1]

Thus we have a sufficient foundation and a sufficient primary cause for the further natural evolution

[1] This does not exclude the possibility of adaptation and selection in the organisms, for the inorganic constellations of atoms in the organisms afford the most varied dispositions as a foundation for the vital processes.

of the whole inorganic world—and this to me appears to be a very reasonable view to take.

4. In order to account for the origin of the first organisms, the theistic theory of life presupposes a so-called act of creation to have taken place. I say 'a so-called act,' for the fact that the first organisms were produced from inorganic matter is intimately connected with this theory. It does not involve creation out of nothing, as the creation of matter does; it is a production of organic bodies out of pre-existent inorganic matter. Should science be in a position to prove that *spontaneous* generation was actually possible, and that living beings could proceed spontaneously from inorganic matter, theism would at once surrender this fourth postulate, for it is merely conditional and not essential to the Christian theory of the universe; in fact, science has forced it upon us, in proving to us by biological facts the impossibility of spontaneous generation. For this reason philosophy also denies its possibility, and this denial necessitates the assumption of a higher cause, of some particular action upon matter on the part of the Creator, in order to explain the origin of the first organisms.

5. The *earliest laws of evolution* were laid down for the organic world at the production of the first organisms. These laws are often misrepresented, as if they were little sprites hovering over the atoms, pushing and pulling them in some

mysterious and supernatural way. This is quite a mistaken view, and in accordance with Christian philosophy we ought to think of these laws of evolution in the following way. In the first place, they comprise the chemical and physical properties of the organic elements and the original mechanical constellations of living atoms, as ordained by the Creator at the production of the primitive types. From these constellations arise certain definite *tendencies of evolution*, which may be further influenced by the reciprocal action of other groups of atoms. In the second place, I agree with Driesch, Reinke, and other modern vitalists in thinking that, if we are to attain to a perfect explanation of the phenomena of life from its first manifestation, we must still accept *formal principles* of the type of the Aristotelian *entelechies*. We have not as yet any chemical and physical formula which explains life satisfactorily and finally. Scientists hope that at some future time an explanation may be found, but I believe this hope is unlikely to be realised. What we call life is something quite different from all the material chemical and physical processes which are subordinate to the functions of life. Natural science therefore compels us to assume certain formal principles, which are not makeshifts meddling with material energy, nor do they disturb the permanence of the law of energy; they simply direct the lower energies, quicken to life the atoms hitherto dead by absorbing them into the organism, and, in short, effect *the purpose* of the vital process by action from within. This postulate is eminently reasonable—I personally cannot dispense with it, and should not be able to dispense with it, even if theology did not exist.

6. We now come to a further postulate required by the Christian theory of life, and it is that against which the monists protest most vigorously at the present time, viz. the assumption that *man possesses a spiritual and immortal soul.* Christian philosophy long ago expressed the opinion that beasts also are not mere machines, and when some modern philosophers declared ants and other invertebrates to be reflex machines, advocates of Christian philosophy, basing their arguments on biological facts, proved this view to be untenable.[1]

We cannot dispense with the assumption that beasts possess some psychical activity, but how far does it go? Only as far as the sphere of the senses extends. On the intellectual side the whole psychical activity of beasts is limited to sense perception, to the connection of such perceptions with one another, to memory, and to the modification of earlier forms of activity in accordance with sense experience. This psychical activity brings into action the inborn tendencies and directs them suitably to perform the

[1] Albrecht Bethe, *May we ascribe Psychical Qualities to Ants and Bees?* Bonn, 1898; E. Wasmann, *The Psychical Capabilities of Ants*, Stuttgart, 1899. For further details regarding my argument with Bethe, see the third edition of my book, *Instinct and Intelligence in the Animal Kingdom*, Freiburg i. B., 1905, chap. viii. p. 157 *et seq.*

vital functions. A beast possessing these faculties is plainly not a machine, but still it does not enjoy intellectual life.

It may be boldly stated that much confusion as to the meaning of the expression '*intellectual life*' has been caused by Büchner and Brehm and other leaders of popular psychology. All our sense perceptions taken collectively are regarded as constituting intellectual life, although they do nothing of the kind. In the sense in which the expression occurs in ancient philosophy, intellectual life is only that form of activity which we describe as 'higher,' viz. the exercise of human thought and human will. What characterises human thought is the fact that man possesses the power to form concepts, and to deduce from them general conclusions, and to raise himself by the aid of his reason above all particular phenomena. On this power depend all the art, science, and religion of mankind, which are not found among beasts, although there are some trifling resemblances to them, which have been exaggerated until they amount to real equality. If we wish to be consistent, we shall require to have a special principle underlying this intellectual activity, which distinguishes man above all the rest of nature, and this principle must be a simple, intellectual being. This soul is not, however, shut up in the human body as in a prison, but with the human body it forms one complete being and substance; hence, in addition to the higher intellectual faculties, it possesses others belonging to the senses, which correspond to those of beasts. In man the one soul is capable of all the activities which a beast can exercise, but in addition it raises itself to the higher functions of the will and intellect, and thus it towers above the sphere of animal life. It is because of this essential superiority of man, in respect of his spiritual activity, to what is animal and material, that we are forced to assume the existence of *a simple, spiritual soul* in man, a soul which continues to exist after death, although it can obviously no longer exercise its lower functions, when once it is separated from the body.[1]

Having made these preliminary statements, I can proceed to a short comparison between the theistic and the monistic doctrines of evolution.

In considering the theistic doctrine we are confronted by the problem of creation. This problem is not contrary to reason, as the finite existence must have a finite beginning in an infinite existence which alone has in itself the ground of its being. This is perfectly reasonable, even if we cannot imagine how it came to pass. Here therefore we have one problem, and if it is solved in the manner approved by the Christian theory of life, it enables us to solve all other problems in logical sequence. But monism presents us not with one problem but with *thousands*, all independent of one another, and

[1] In Part II. will be found some critical remarks upon the objections raised against this postulate by Dr. Juliusburger in his speech during the evening discussion.

all incapable of solution. The first is *the indestructibility of matter* and its laws. If this is considered from the philosophical point of view it at once appears bewildering. It is not possible to connect the idea of endless existence with matter, because in its very nature it is subject to change, and only what is not subject to change can be everlasting.

Moreover, the problem of *the origin of motion in the universe* is insoluble, as Du Bois-Reymond pointed out many years ago. If everlasting matter was of itself in a state of inertia, whence originated its motion? If, on the other hand, matter was of itself in ceaseless motion, how has it come to pass that we have not yet reached an equilibrium of all forms of energy, and the rigidity of death throughout the universe.[1]

Everywhere we meet with innumerable difficulties. As to the problem regarding the laying down of *the laws of nature*, which cannot have taken place spontaneously, it has been asserted that these laws developed accidentally out of the original chaos—an assertion that even Darwin's followers have rejected as incompatible with reason. *Just as only a mind capable of thought can form any conception of the order of the universe, so only a mind capable of thought can in the first instance have produced that order.*

[1] Cf. R. Stölzle, 'Has Laplace's Theory of the Formation of the Universe an Atheistical Tendency?' (*Natur und Kultur*, iv. Nos. 9, 10, 11, 13).

Let us now consider more closely *the interior laws of evolution* in the organic world. The monists maintain such 'inexplicable' laws to be superfluous. But if we assume no more than that the living substance possesses a capacity of reaction, when exposed to external stimulus, we are at once confronted with the principle of *expediency*, which cannot be further explained (by monists), because the *idea of purpose or design* is inseparable from it.

I wish to emphasise this statement. In protoplasm the faculty of reacting conformably to an end implies the existence of an intrinsic law of evolution. Such a law is—*the interior laws of evolution are inseparable from the idea of capacity on the part of the protoplasm to react in accordance with expediency.* These laws are absolutely indispensable, because we cannot imagine a living protoplasm devoid of the vital purpose expressed in the processes of growth, nutrition, and propagation.

It would, however, be a mistake to regard the interior laws of evolution, which the theistic theory of life assumes as the chief principle underlying the evolution of the organic world, as a sort of clockwork, wound up once for all, and left to run down. We must not assume the existence of any 'pre-established harmony' between an organism and the world around it; no, it is rather reciprocal action and the disposition to reciprocal action, which allows the interior and exterior factors in evolution to work together. When people speak of the sensitiveness of the protoplasm to stimulus, of its capacity for reaction under external influences, they mean something identical with the interior laws of evolution. Exterior influences determine their direction, and they are rendered permanent by transmission, and thus fresh lines of evolution are constantly coming into being, which, though of a more specialised kind, nevertheless rest primarily on the same interior basis whence they proceed. We cannot deny the existence of intrinsic laws of evolution, as do many of the supporters of Darwin's theory of Natural Selection, without entangling ourselves in a web of contradictions.

I have nothing to add to what has previously been said, in order to justify the assumption that man possesses an *intellectual soul.* We cannot avoid this postulate, and the noblest intellects of every nation, ever since the beginning of the first period of intellectual culture in the world, have stoutly maintained the existence of an intellectual and immortal soul in man, and I believe they will continue to maintain it in the future.[1]

I must acknowledge frankly that, as a scientist, I am by no means ashamed of being an adherent of the theistic theory of life, *because I regard it as the only correct one.* This avowal is not directed against the supporters of monism, but against monism itself. I believe that the arguments

[1] Cf. K. Kneller, *Christianity and the Advocates of Modern Natural Science*, Freiburg i. Br., 1904.

brought forward by monists against the Christian theory of life are based for the most part on misunderstandings.[1]

Darwinism and the Doctrine of Evolution.

I now come to *the comparison of Darwinism with the theory of evolution.* This comparison is really superfluous on this occasion, when so many students of zoology and other branches of natural science are present, for all that I have to say has been long known to zoologists, but unfortunately such is not the case among the general public. At the meeting of German scientists at Aix-la-Chapelle in September, 1900, Oskar Hertwig rightly maintained, in agreement with Huxley, that the doctrine of evolution would remain unaffected if the Darwinian theory were given up. In other words, *Darwinism and the doctrine of evolution are not equivalent ideas.* The latter, which is wider and more general, connotes the doctrine of the derivation of *all* forms of life from earlier and simpler forms, whereas Darwinism deals with the origin of the organic species *by way only of natural selection,*

[1] The opinion expressed once by Linnæus in the following well-known words is undoubtedly a lofty one and worthy of a true student of Nature: 'Deum sempiternum, immensum, omniscium, omnipotentem expergefactus a tergo transeuntem vidi et obstupui. Legi aliquot eius vestigia per creata rerum, in quibus omnibus, etiam minimis ut fere nullis, quae vis, quanta sapientia, quam inextricabilis perfectio! . . . Numen esse credi par est, aeternum, immensum, neque genitum, neque creatum' (*Systema Naturæ*, 13th ed., 1789, p. 3).

and is therefore a special branch of the doctrine of evolution. This is Darwinism in the historical sense of the word, although Darwin himself was not such an extreme Darwinist as many of his followers have been, for he recognised other concurrent factors in evolution, whilst laying the greatest stress upon natural selection.[1]

The word *Darwinism* has various significations, and consequently a great confusion of ideas with regard to it prevails among those who are not specialists, and this confusion is increased by certain works on popular science which are particularly obscure in this respect. I am referring especially to France's work on the Further Evolution of Darwinism, which has been published among Breitenbach's collection of writings bearing on Darwinism. The author speaks of the 'Further Evolution' of Darwinism, and yet he reduces the value of the principle of Natural Selection to a minimum, by acknowledging that it is an unimportant subsidiary factor in evolution. It is time for such confusion of ideas to cease. Darwinism as such is Darwin's theory of selection and nothing else.[2]

For the last forty years, in all his popular works

[1] *E.g.* direct adaptation, correlation, compensation, etc. Cf. *Origin of Species*, chap. v.

[2] The principle of selection was originally formulated by Wallace (1858), to whom priority is ascribed in this respect. However, as Darwin very soon afterwards (1859) in his *Origin of Species* worked out the principle for the first time in its universally accepted form, and absolutely independently, the theory of selection received the name of Darwinism.

Haeckel has been confusing Darwinism with the theory of evolution, and his reason for so doing, as he stated in his address on 'Monism as a Bond of Union between Religion and Science,' is that Darwin's Theory of Selection supplied him with the only means of explaining orderly action in nature without assuming a designing or ordering Creator. But in the first of his Berlin lectures in 1905 he expresses another view, and on page 20 of his book entitled *The Struggle concerning the Idea of Evolution* we read:

> 'It is the theory of Selection, which supplies a causal explanation of the processes attending the formation of species, *that should, strictly speaking, be described as Darwinism.* How far this theory of selection can be justified, and how far it is liable to modification by other theories, such, for instance, as Weismann's Germ-plasm theory, or de Vries's theory of mutations, we cannot discuss at the present moment.'

He did not discuss it in his subsequent lectures, and I can only account for this by supposing that Haeckel has finally seen that if we do not limit Darwinism to the theory of selection, we shall have to let the name go altogether—and therefore he prefers to say no more on the subject. But the word Darwinism is still commonly used in the earlier sense, and men's ideas continue to be confused. I have, I think, said enough to prove that we are perfectly justified in demanding a clear distinction between *Darwinism in the narrower sense* and *the theory of evolution.*

The theory of Selection, or Darwinism in the narrower sense, is briefly as follows:—Just as a man, who breeds domestic animals, chooses from the different varieties definite individuals possessing definite qualities, in order to breed from their intercourse a new race possessing these qualities, so is there a similar process in nature, not motived by any purpose. The hypothesis, on which this theory rests, is that the organic species are subject to change, working in various directions in an undefined and unlimited manner. If under certain conditions some varieties are produced, which accommodate themselves better than others to the circumstances of their life, these will triumph in the struggle for existence, and the others will be eliminated. The victors will eventually transmit their qualities to their descendants, and by this transmission the qualities will become more and more prominent, until a new variety, a new race, a new species, etc., has been developed. This underlying thought of the Darwinian theory is partially correct, and much may be said in its favour: I do not reject it, but its range is not so wide as it is often believed to be.

In the wider sense, Darwinism is the name given in popular circles to the generalisation of Darwin's theory of selection, and its extension to a 'Darwinian theory of the universe.' This is identical with the monistic theory in the form of Haeckelism; according to it the whole world has come into existence without a creator and through merely mechanical causes.

The third way in which the word Darwinism is used, popularly, is to designate the application to man of the Darwinian theory of selection. Man is assumed to be the animal most highly bred in the course of the struggle for existence, and nothing else.

Fourthly and lastly, the name Darwinism is applied in a general way to *the theory of evolution*, as I remarked before. This confusion of ideas has done much harm in many ways. If, for instance, a serious student, engaged in scientific research, finds in his special department what he regards as evidence of the development of species, he is at once called a Darwinist, and as such is assailed by another party. In the same way, on the other hand, the advance of the theory of evolution as a scientific hypothesis and theory is quite wrongly appropriated as an outcome of Darwinism, as Haeckel especially has done. This explains the great applause which Darwinism has received in the widest circles and down even to the lowest classes.[1]

Let us now attempt to give a short *criticism* of these various ideas of Darwinism.

The theory of selection has recently been very adversely criticised. Certain scientists refuse altogether to accept it, and some men of note have

[1] See also *Modern Biology*, p. 265 *et seq.*

expressed very unfavourable opinions regarding it. Driesch, for instance, says that Darwinism was one of the great delusions of the nineteenth century, which was completely taken in by it, and that Plate's eulogy of Darwinism sounded to him like a funeral oration. This unfavourable view concerns, however, only the extreme form of Darwinism, which seeks to explain everything exclusively by selection. My own experience, gained in the course of research work in my special department, shows natural selection to be indispensable as a subsidiary factor, but only a *factor*—the *interior causes of evolution* remain always the chief point to consider, for they produce the beneficial modifications, and so are of greater importance than external circumstances, for these only eliminate the modifications which are not beneficial in the struggle for existence.

We ought, moreover, always to consider the various principles of evolution *collectively*, and selection is only one of them, and, moreover, a subordinate one, which in its very nature bears a negative character—for it only weeds out. It is, of course, possible, as Professor Plate rightly pointed out in an excellent dissertation on Darwin's principle of selection (2nd ed. p. 187), that the result of this negative selection may in many cases be positive, for by means of it definite tendencies to evolution may be logically furthered (Orthoselection), and so something positive is produced. The action of natural selection is, however, always essentially negative; it is the *survival of the fittest,* and *the underlying reason for the presence of the fittest* must be sought elsewhere, ultimately in the *interior laws regulating the evolution of organic life.*

In this connection the theory of direct adaptation, put forward by Lamarck and Nägeli is of great use, although it is only another way of stating the beneficial capacity for reaction to external stimulus which organisms possess. I believe that it is quite impossible to avoid regarding the interior laws of evolution as of *primary importance,* although I gladly allow that it is difficult to work with unknown causes. The exterior impulses governing evolution, which are presented to us by means of plain examples in Darwin's theory of selection, are very attractive. As I showed you yesterday in speaking of the inquilines among ants and termites, it is easy to give many interesting examples of the exterior conditions affecting evolution, but the exterior factors could not be effective if they did not correspond with the interior factors. *The co-operation of the interior and exterior factors is absolutely necessary for any beneficial adaptation.* If we do not as yet know the interior factors of evolution, that is a defect due to our imperfect scientific knowledge; we have only made a very humble beginning towards investigating *the causes* of organic evolution, but no one can lay the blame on the theory of evolution, which is still in its infancy.

Possibly in a hundred years' time a new theory will be discovered, based on Boveri's hypothesis of the individuality of the chromosomes, according to which, by means of definite modifications of the chromosomes in the germ cells, corresponding modifications in the process of evolution can be accounted for in the organisation of the germ-plasm—not on the lines of Weismann's theory of determinants, but rather perhaps on those of O. Hertwig's theory of biogenesis. A theory of this kind would show that considerable progress had been made. It is in this way that I imagine the interior causes of evolution will be discovered. I do not think of them as of mysterious sprites hovering over the waters, a view the monists impute to me. I have never put forward explanations of this kind, but they have been attributed to me, in order to facilitate the elimination of the interior laws of evolution.

A few words must be devoted to Weismann's views on the subject of natural cultivation. Several years ago, Professor August Weismann, a very skilful zoologist, laid great stress on the *all-importance of natural selection*, by means of which he, at that time, thought every phenomenon could be explained, in conjunction with Darwin's theory. But more recently Weismann has abandoned his extreme views ; he no longer gives the chief place to natural selection. Since 1895 he has very cleverly thought out the theory of *germinal selection*, but if we carefully examine the meaning of obscure expressions such as the 'vital affinities' of the biophores, etc., we finally discover that they conceal the ideas of purpose or design, of adaptiveness, and of capacity of reaction under external stimulus; in short, these expressions include all the interior laws of evolution, which were supposed to be set aside. I believe therefore that Weismann's Neo-Darwinism actually serves to prove that Darwin's theory of selection is untenable, if it is carried too far.

Not long ago I studied very carefully the new edition of Professor Plate's work on Darwin's principle of selection. It is perhaps the best work that has recently been written in support of this theory. What interested me most in reading it, was to see how the most faithful adherents of the principle of selection have at last begun to recognise its limitations. Professor Plate states them quite calmly and fairly, but on the other hand he seems to me to go a little too far in emphasising the merits of this principle. I do not agree with him on several points, especially I think him wrong in rejecting the interior laws of evolution and teleology; but we undoubtedly owe Professor Plate a debt of gratitude for his excellent criticism of Darwin's principle of selection.

I need not waste time in discussing the so-called *Darwinian theory of the universe*, because it is identical with that realistic and monistic theory with which I dealt at the beginning of this lecture. As to the application to man of Darwin's theory of selection, let it suffice to say that it is untenable because it regards man too exclusively as an animal; but the application to man of the theory of evolution will be the subject of my third lecture on Sunday next.

I may sum up shortly the results of my examination into the scientific value of Darwinism in the following way:—

Darwin's theory of selection is indispensable as a *subsidiary factor* in the theory of evolution; but its value is subordinate and varies very greatly, according to the class of phenomena with which we are concerned. For instance, among the guests or inquilines of the ants and termites we recognised yesterday a hostile type, calculated to resist attacks; a mimetic type, in which the guests deceive their host by their close resemblance to them; and, lastly, there is the type of true guests. The theory of selection, when applied to these three types, has quite different results. It is most important in the case of the hostile type, somewhat less so in the case of the mimetic type, and least of all in the third or symphilic type, in which we find the principal factor to be *amicable selection*, which is not only different from natural selection, but, from a certain stage of development onwards, is in antagonism to it and prevails over it.[1]

[1] Cf. *Modern Biology*, pp. 338, 345, 384.

An example will show this. The blood-red robber-ant, in entertaining as guest the *Lomechusa*, is supporting its worst enemy, and it does so through an instinct leading to the destruction of its own species. We see here an instinct which cannot possibly have been the result of natural selection, for the guest is harmful from the moment when it deposits its larva to be brought up in the ant's nest. I believe that in this case amicable selection has triumphed over natural selection, but I am far from assuming that the theory of selection has been equally ineffective in all cases. In fact, many instances can be given in favour of this theory, but it is effective only when interior adaptivity can be assumed on the part of the organism. This quality is absolutely necessary.

In the first part of this lecture I pointed out the contrast between the Christian theistic theory of the universe on the one hand, and the monistic view on the other, which denies the existence of a personal God and Creator. In many scientific circles there is an absolute *theophobia*, a dread of the Creator. I can only regret this, because I believe that it is due chiefly to a defective knowledge of Christian philosophy and theology. The study of one thorough text-book, such as Gutberlet's *Theodicee*, would suffice to give a clearer idea of the significance and true meaning of the Christian conception of God.

Finally, I should like to call as witness, in support of the theistic view, one who certainly cannot be suspected of being a Jesuit. Charles Darwin had not that morbid fear of a creator which seems to dominate many of his followers.[1]

At the end of his great work on the *Origin of Species* occurs the following beautiful passage, which stands unchanged in the seventh German edition published after his death:—

'*There is grandeur in this view of life, with its several powers, having been originally breathed by the Creator into a few forms or into one; and that, whilst this planet has gone cycling on according to the fixed law of gravity, from so simple a beginning endless forms, most beautiful and most wonderful, have been and are being evolved.*'[2]

I think that after these words I myself, as a scientist, need not apologise for being an advocate of the theistic theory of life.[3]

[1] It is well known that in his later years Darwin inclined to Agnosticism. The fact, however, that in the subsequent editions of his *Origin of Species* he did not alter the words quoted here, shows plainly that he was in no way ashamed of his earlier theistic opinions.

[2] *Origin of Species*, 6th ed., 1888, vol. ii. p. 305.

[3] Cf. also the lecture delivered by Professor Reinke of Kiel on Natural Science and Religion, printed in the *Propyläen* of March 13, 1907, No. 24.

LECTURE III

(*February* 17, 1907)

THE DESCENT OF MAN

THE question whence man has come and whither he is going has always marked a point where opinions diverge. Man is formed of dust and returns to dust—and this has always been maintained by materialists to be his whole destiny. The idealism of every age of Christianity has, on the contrary, asserted that *this is not his whole destiny*. A 'spark of the divine spirit' dwells within the mortal body, and just as the *origin* of man is to be sought in God's creative power, so God is the ultimate *end* of his life, and the human spirit can find perfect happiness nowhere save in the knowledge and love of God, in an everlasting life after death.

What has natural science to say on this subject? Man is undoubtedly formed of dust and returns to dust, if we consider merely the lower or animal side of a human being. Modern biology teaches us that man, like other vertebrate animals, is developed from a diminutive cell, and it shows us further that in man, as in beasts, the *germ cells* are the chief agents in heredity, they are the one constant element in the bodily history of humanity, whilst the individual comes into existence and passes away. But biology has considered the question from its point of view in *only one, and that a material, aspect. The other, i.e. spiritual, aspect* of the same problem falls outside its scope, and the results of biological investigation do not touch the existence in man of a soul created by God, and destined, after the death of the body, to return to God.

Similar remarks will apply to the *hypothetical history of the human race.* It may on its material side originate in the dust of the earth, it may during its whole course of existence be inseparable from the dust of the earth, and it may finally return to the dust of the earth—and yet in all this there is nothing derogatory to the dignity which man possesses as *God's likeness,* in virtue of his *spiritual soul,* there is nothing at variance with his being originally of divine creation, and with his being destined ultimately for a divine goal.

Every atom in the human body had its primary origin in a creative act of God at the first formation of matter, although millions of years of cosmic development were to elapse, before it became a living part of a human body; and, in just the same way, we might imagine a hypothetical history of humanity, governed by the laws of natural development, which God impressed upon the first cells at the moment when life originated.

In accordance with this purely speculative supposition, man would have become man completely only when the organised matter had so far developed through natural causes, as to be capable of being animated with a human soul. The creation of the first human soul marks the *real creation of the human race,* although we might assume that a natural development lasting millions of years had preceded it.

These are, it is true, only attractive possibilities, the outcome of bold speculation, but I have referred to them here in order to prove to you that, if ever science is able to demonstrate to us the natural development of man from an ancestry resembling beasts, *the divine origin and the divine end of humanity will nevertheless remain unassailed and firmly established as before.*

Let us now return to the dry, serious, and scientific subject whence our speculations have caused us to stray.

We here reach the most delicate and most important question, which to some extent is the rock of offence in the whole doctrine of evolution, viz.: '*May this theory be applied to man, and if so, in what degree?*' I wish to state definitely, before discussing the matter, that we are not concerned with the application to man of Darwin's theory of evolution, for I showed in my last lecture that I was unable to accept that. We may apply the theory of evolution to man, and still have as foundation the principles of Christian philosophy and of the Christian theory of life.

But even in this case we are confronted with the question: 'If we take as our basis the Christian theory of the world, viz. creation connected with evolution, (inasmuch as the Creator has created a world capable of development),—can we, or can we not, include the formation of man in such an evolutionary process?'

The question is a very difficult one, chiefly from the fact that it is not a simple, but *a very complicated, problem.* Not only are the natural sciences concerned, but theology asserts her right to decide in what way man came into being; and within the domain of the natural sciences, psychology, the science of the mind, claims also to express an opinion on the subject. In short, the question that we have to discuss this evening *is not a purely zoological one,* and we must do our best, as far as possible, to do justice to all the various aspects of it, and not to confuse them with one another.

We often hear allusions made to the so-called *zoological evidence* for the descent of man from beasts. If the descent of man from beasts were proved to evidence by zoology, then to zoology would belong the right, predominantly or exclusively, to decide the question of the origin of man; and further, zoology should already have brought forward definite proofs—not mere general possibilities, but real evidence, with regard to which scientists would be in agreement. We have now to consider how far this is the case.

Is it the sphere of *zoology alone* to express an opinion regarding the origin of man? Is this science alone competent to do so? No, it is not—although it would be competent, if man were merely an animal and nothing more. Were this the case, we might well ask: 'Whence can we suppose man to have come, if not from a tertiary mammal? Surely he did not fall from the skies?' But as a matter of fact man is not merely an animal. Whoever recognises an essential difference between man and beast, and regards the intellectual soul of man as his most important part, will acknowledge that in investigating the descent and origin of man, the chief question is: 'Whence comes his *higher* part?' not: 'Whence comes his *lower* part?' Therefore I believe myself justified in saying that psychology, and not zoology, is of chief importance among the natural sciences, when they are called upon to account for the origin of man. Now psychology tells us—I am speaking especially of psychology as a department of Christian philosophy—that the soul of man is not only essentially different from the soul of an animal, but is a *simple spiritual being.* Now such a being cannot in its very nature develop out of anything else—it can come into existence only by way of creation. Therefore the *soul* of man cannot owe its origin to evolution.

What follows then with regard to evolution applied to man as a whole ?

It follows that man as a whole cannot, as such, have been produced from any animal form, by way of purely natural development. We have still to consider the subsidiary question: 'Is man with respect to his body related to the animal kingdom by way of descent ?'

Before discussing this question from a scientific point of view, we must refer shortly to the theological side of the problem.

It is well known to you all that the Mosaic and the Christian religions, following the biblical account of the Creation, teach that God created man in a way peculiar to man. We read that 'God formed man of the slime of the earth, and breathed into his face the breath of life.' It is plain that 'breathing in the soul' is only a symbolical expression for creating the soul of man. What are we to say about the formation of the human body from earth ? The Church has not promulgated any definite decision as to the nature of the substance employed by God in the creation of man. Theologians, however, following constant tradition and the opinions of the ordinary teaching authority in the Church, have consistently maintained that the human body was formed of inanimate matter. Perhaps this is all that need be said on the theological side of the question.

Let us pass now to the scientific side of it—and

at the outset let me emphasise the fact that zoology and its attendant sciences, as long as they limit themselves to their proper sphere of investigation, are perfectly free to discuss the origin of man. The assured results of theology need serve them only as an external standard, for one truth cannot contradict another. If science reveals some undoubted truth, theologians will accept it. I can vouch for the accuracy of this statement.

What *scientific proofs* are there of the descent of man from beasts, as far as his body is concerned?

We must distinguish two chief classes of proofs, the *zoological* and the *palæontological.*

I must try to express myself in the most concise manner possible in dealing with the enormous mass of facts presented to us for consideration.

1. Under the heading of *zoology,* we must first refer to *comparative morphology,* as far as it is concerned with the question. It tells us that the human body is that of the most highly developed mammal; and it shows us many resemblances between man and other mammals in the formation of the skeleton, of certain organs and of the nervous system. It would not occur to any scientist to deny that these facts lend a certain amount of probability to the theory of man's descent from beasts, but, on the other hand, we must not overlook the *various differences,* which comparative morphology reveals as distinguishing man and beast, and especially as marking off man from the higher

or anthropoid apes. Ranke has stated these differences in an impartial and objective manner in his work entitled *Der Mensch* (2 vols.). It is impossible to consider them all in detail here, and I can mention only a few points.

Walkhoff has examined the femoral bones of man and of the ape, and declares that any Röntgen-ray photograph of a frontal section of the bone, or even of a part of it, would enable an expert to decide at once whether it had belonged to a man or to an ape.[1] The upright walk of a man requires a totally different arrangement of the fibres of the bones, from that which exists in an ape. The upright walk, however, is connected with the formation of the cranium. The occipital foramen in man occupies a different position from that which it has in other vertebrates, because the relative development of his cranium is much greater, and it is greater particularly in the region of the brain, and far exceeds the face in size. The reverse is true of apes. And why? Because man needs a more complete development of the brain to serve as the indirect instrument of his intellectual life. On account of the height of his intellectual position, his imagination and power of visualisation, which are directly dependent upon the brain, must be far more highly developed than they are in other animals,

[1] It would be very interesting to examine the femur of the Pithecanthropus by means of the Röntgen rays, according to Walkhoff's method, in order to determine whether the creature really walked upright or not.

as they have to do the preliminary work which serves as the basis of intellectual activity, properly so called. We may therefore say that *all bodily differences between man and beast are ultimately a result, or rather a function, of the mental difference between them.*

The lecturer made use of a diagram, placed at his disposal by Professor Plate of the Agricultural College, to draw attention to the characteristics of the anthropoid apes (chimpanzee and gorilla), their prominent jaws and the extraordinary development of their arms.

A few lantern slides brought out the points of difference between a human skeleton and that of one of the higher apes: the much longer extremities of the ape—his foot, which is really a hand, and does not resemble a human foot. The human skull showed much greater development in the region of the brain. Ranke is quite right in describing man as a brain-animal, as in a human skull the part containing the brain has a much greater capacity than all the other parts of the head. The lecturer displayed the skull of an orang-outang by way of contrast. The face is enormous, but the cranium is comparatively very small. The powerful jaw with its huge teeth reveals the animal nature, and it is obvious that in this case a mere struggle for existence is the chief object.

The lecturer continued:

In showing you these diagrams my intention was to draw your attention to a few salient points of difference between the human skeleton and that of an ape; they are so great that it is impossible simply to ignore them, and we can bridge over the chasm separating the crania of men and apes respectively, only by making assumptions that are not justified by facts. To the present day no connecting link has been discovered.[1]

Zoology, or rather the *comparative history of individual evolution*, has furnished another series of arguments in favour of the descent of man from beasts. This *biogenetic principle* was laid down first by Fritz Müller and developed later by Ernst Haeckel. According to it the development of the individual is only an abbreviated and partially modified reproduction of the development of the race. Haeckel worked out the application of this principle to man in great detail, and tried to prove that man in his embryonic growth passes through twenty-two—later on the number was raised to thirty—stages of development, corresponding with the same number of stages of ancestors, some of which answer to still existing animal forms, but others are purely imaginary and postulated by Haeckel for the sake of his theory. This argument attracted much attention and found many to support it in popular circles. People scarcely ventured to doubt that man, in his individual

[1] What are believed to be the oldest human, or quasi-human, remains ever discovered have been unearthed near Chapelle-aux-Saints, in the Department of the Corrèze, and acquired for £60 by the Paris Museum of Natural History. M. Perrier, director of that institution, in a communication to the Academy of Sciences, assigns the remains to the Pleistocene or Glacial Period. From description they appear to be the long-sought missing link, being neither man nor ape, but having characteristics of both. The skull more resembles that of a human being, but the shape of the limbs indicates that the creature walked on all-fours rather than erect. In close juxtaposition to the skeleton were found the teeth of a rhinoceros.—*Reuter.*

development, reproduced a number of ancestral forms in succession, and that this constituted a conclusive proof of his descent from beasts. I believe, however, that recent investigations in comparative morphology and in evolution justify very serious doubts as to the accuracy of the biogenetic principle in general, and especially as to its application to man.

The first point—*the accuracy of the biogenetic principle in general*—has been discussed by Karl Ernst v. Baer, and its true value has been ascertained.[1]

Quite recently it has been critically examined by Oskar Hertwig in his *General Biology*, and in the last chapter of his *Handbook of Comparative and Experimental Evolution of Vertebrates.* According to the most valuable statements made by Hertwig, the evolution of the individual is not a *repetition* of that of the race, but, assuming the principle of evolution, we must regard it as a *continuation* of the development of the race. As this process continues, the corresponding new generation must advance somewhat further than its immediate predecessor; there is no simple repetition of the evolution of the race.[2]

[1] See also J. Reinke, *The Laminariaceæ, and Haeckel's Biogenetic principle*, Kiel, 1903.

[2] Haeckel himself shows that he recognises this fact, for in the development of the individual he distinguishes Palingenesis, or repetition of ancestral forms, and Cænogenesis, or new formations. The latter, in fact, so often outweighs the former in importance as to overthrow the 'prin-

The apparent repetition of many previous stages of development is accounted for by the fact, that it is essential to the very nature of evolution to advance from what is simple to what is complex. The more highly any animal is organised, the more stages of development must it pass through, before reaching the complex final stage, and it is quite in accordance with nature that the previous transitional stages, being simpler, should resemble the final stages of other animals, which have remained stationary at a lower degree of organisation. This constitutes no proof that the human race has passed through all these stages, but it only shows that the evolution of the individual goes on from the first subdivision of the impregnated ovum, through various stages, until the final form of the perfect organism is reached.

I maintain therefore that *we cannot accept the biogenetic principle in its entirety, nor can we sanction its application to man in order to prove his descent from beasts.*

I shall be met, no doubt, with the objection that there are stages in the development of the individual human being, for which no explanation is possible, unless we regard them as repetitions of an earlier race-evolution. The most important of them in this connection are the so-called *branchial arches and clefts* of the human embryo.

ciple' altogether. On this subject cf. my *Biology*, chap. ii. pp. 457, etc. See also the remarks on Dr. Schmidt-Jena's speech in Part II. of the present work.

They occur to the number of four and three respectively in all mammals and in human beings.

In the case of fishes they develop eventually into real branchial arches and real branchial clefts. If we consider what becomes of them in the case of the higher vertebrates and man, we find that the first branchial arch becomes the mouth, and the first branchial cleft the exterior ear; the others either undergo involution, or they form various other organs, the bones of the inner ear, etc.

If we consider the matter in a calm and dispassionate way, we arrive at the conclusion that, in the case of the higher vertebrates and man, these so-called branchial arches and clefts are merely curves and folds of the pharynx, which are quite unimportant in themselves, and eventually develop into something bearing no resemblance to real branchial arches and clefts. They are, in fact, simply *pharyngeal arches and clefts.* But in the case of fishes, to whose existence gills are essential, a similar arrangement develops into real gills, and so, with regard to them alone, it is correct to speak of real branchial arches and clefts as existing in the embryo. I can by no means see that these facts constitute an argument in favour of a theory that all mammals, and man in particular, have passed through a fish-like stage of being, nor is there any logical support for such a theory.

It would be a mistake, however, to assume that I simply reject the biogenetic principle. If it only means that certain cases occur, in which the individual development of a creature throws light upon the hypothetical development of the race, I am willing to accept the principle, but then it ceases to be a general law. It is an undeniable fact that, among both the higher and the lower animals, instances occur of stages of individual development, which can be explained only by regarding them as temporary traces of a previous stage of development, which was permanently impressed upon certain ancestors. As an example of this, I may refer to the teeth which the embryos of the whalebone-whale still possess, although subsequently they degenerate into whalebone. Geoffroy Saint-Hilaire first observed this fact early in last century, and Kükenthal has confirmed his statement. If we compare with it the further fact that geology has ascertained, viz. that the whalebone-whale only in the tertiary period succeeded the toothed whale, which may be regarded as its probable ancestor, the conclusion is obvious. The whalebone-whale is descended from an older toothed whale, and the reason why, in the development of the individual whalebone-whale, there is a stage at which teeth appear, lies in the fact that the ancestors of the present whales passed through this stage of development, and it remains up to a certain definite point in the growth of the embryo.

Something similar occurs in the case of the Termitoxenia, a very small fly that lives with the white ants. You saw a diagram of it during my first lecture. It presents the peculiar feature of having for a short time, whilst it is passing through the stenogastric stage as a full-grown insect, genuine veined wings in the still cuticular appendices to the thorax.

I could scarcely believe my eyes, when I noticed this for the first time in my series of sections. Subsequently these little hooked appendages to the thorax grow into horns, and serve as organs of touch and exudation, and enable the fly to balance itself, and no trace of likeness to wings remains. Probably we have here a certain amount of reproduction of the growth of some ancestors. Originally these appendages developed into true wings, now the rudimentary wings change into other organs, serving quite another purpose; but as this change is not so remote, we still find a temporary reproduction of the former winged stage, when real wings appear for a short time. There is no trace of wings in the Termitomyia, which has departed further from the diptera type. In it the hooked appendages on the thorax appear at once, without any reproduction of an intermediate stage.[1]

I might refer to a number of similar instances, but what has been said will suffice to show that

[1] Cf. 'Die Thorakalanhänge der Termitoxeniidae,' *Proceedings of the German Zoological Society*, 1903, pp. 113-120, and Plates II. and III. Also *Modern Biology*, pp. 390-392.

there are really cases, in which the evolution of the individual gives us a clear indication where to seek the ancestors of the race. Nevertheless, if we are to explain such a stage of evolution as being a repetition of some hypothetical stage in the life of its ancestors, this explanation must be the only possible one—and it is my opinion that there is no such stage in the ontogeny of man; and therefore I maintain that the individual development of man supplies us with no evidence of his descent from beasts, which can be regarded as conclusive from the scientific point of view.

Zoology has supplied a third argument which is used by those who uphold the theory of the descent of man from beasts. They refer to the existence of certain *rudimentary organs*, which at one time served some definite end, but later degenerated as useless, and now remain in an altered or diminished condition. We must bear in mind that it has been a common mistake to describe as 'rudimentary' any organs of which the use was unknown. Especially in the case of the human body, it has repeatedly happened that organs described as rudimentary have been found to fulfil some definite function of biological importance. I may remind you of the thyroid gland, the thymus gland, and the pineal gland. With regard to the last, Cyon proved that it was important for the maintenance of equilibrium. But there undoubtedly are certain rudimentary organs for which we cannot account in such a fashion. As such, up to the present time, we reckon the *processus vermiformis*, which so often causes appendicitis. We are told that the ancestors of man had a much longer intestine than we now possess, and the vermiform appendix of the cæcum is what remains of it. It is possible that, within the genus Man, in course of time gradual modification of some part of the intestine may have taken place, owing to change of food. We know that the herbivorous animals have a much longer intestine than the carnivorous, and the transition from a vegetable to a meat diet may have brought about a shortening of the intestine. There is much to be said in favour of this explanation, and one point seems to me particularly worthy of notice. The peculiar pathological character of the vermiform appendix is perhaps the result of the hypercivilisation, of the over-refined diet, of the present day. It would be interesting to prove by statistics whether morbid manifestations of the appendix do not occur much less frequently among people living more natural lives. Ellenberger has recently published a work on the appendix, but I am inclined to think that the last word has not yet been said on the subject, and it cannot be regarded as supplying any trustworthy evidence of the descent of man from beasts.[1]

[1] See also the remarks on von Hansemann's speech, which forms part of the evening discussion (in Part II.). It would seem that the vermiform appendix is not a rudimentary organ at all.

There are other rudimentary organs of less importance, to which I need only refer; such are the wasted muscles of the ear and face. Our earliest ancestors may have lived under conditions which forced them to use the ear muscles much more than we do. I acknowledge that in many ways it is difficult to account for the rudimentary organs, but I maintain that there is no conclusive evidence of their phylogenetic significance. But possibly some one will reproach me with having failed to mention the chief proof of man's descent from beasts, viz. their blood-relationship. I am now about to discuss this point.

We have to distinguish *two zoological theories*, both put forward by those who regard man as descended from beasts. According to one, *man is directly related to the higher apes*; according to the other, *he is not directly related to them, but only remotely connected through some ancestor*, from whom both men and apes are descended.

The theory of the *direct* relationship between man and the anthropoid apes was upheld by Karl Vogt in his day, and more recently by Haeckel, and by numerous modern zoologists. Selenka regarded it as definitely proved, when he had established the resemblance in the formation of the *Placenta bidiscoidalis*, as it exists in man and in the higher apes.

But other evidence is commonly adduced, which is based upon the resemblance that exists between human blood and that of the higher apes. *In-*

vestigations into the reaction of the blood have been made by Friedenthal, Nuttal, Uhlenhuth, Wassermann, Schütze, and others, in some cases with the express intention of tracing the relationship between man and the higher apes, in others for various purposes. Some years ago Dr. Friedenthal published a work in which he declared, as the result of his researches into the reaction of the blood, that *man was not only descended from apes, but was a genuine ape himself*. This conclusion is deduced from the following facts. It is ascertained that if the blood of one class of vertebrates, especially of mammals, be injected into the veins of other animals, symptoms of disease appear, in consequence of the decomposition of the red corpuscles of one kind of blood by the serum of the other. There is, however, no such result when the two kinds of animals are closely related. Careful experiments have proved that there is a very feeble reaction between human blood and that of apes, and this has led to the inference that man and the anthropoid apes must be closely connected. Conversely, in the reaction of the antiserum, the morbid effect is most marked in the case of animals most closely related.

Let us apply a critical standard to these experiments and deductions. The experiments are extremely ingenious, and are not only of great use in forensic medicine, but they throw a very interesting light upon the relationship existing between various species. But we are not justified in regarding *a chemical and physiological resemblance between two kinds of blood as constituting a blood-relationship in the sense of having a common origin.* Let us assume that there is a resemblance between the blood of apes and that of men. This would prove that the same kind of likeness exists in the blood of men and apes, as in their skeletons and other organs. But similarity of blood does not imply blood-relationship, such as exists between cousins and kinsfolk. Rössle has recently brought out an interesting work on this subject (in the *Biologisches Zentralblatt* for 1905, Nos. 11 and 12). He is of opinion that the blood-reaction only enables us to say that one animal is more closely related to another than to a third, but it does not show how closely any two animals are related. From his point of view therefore the blood-reaction between man and the higher apes does not justify the conclusion that they are closely related, still less that man ought to be classed with the higher apes. He also insists upon the fact that the chemical composition of the fluids of the body, such as the blood, is no more constant than, for instance, the formation of the skeleton, therefore evidence based on resemblance of the blood is no more trustworthy, in support of a common descent, than that based on similarities of the skeleton and other morphological resemblances. In fact, it has been ascertained that in many cases similarity in the blood does not correspond with morphological resemblance, and the blood-reaction points to a close relation between creatures that are morphologically far apart. It would seem that we cannot make much use of evidence derived from similarity of blood, if comparative morphology arrives at contradictory results.

Moreover, recent investigations by Uhlenhuth and Friedenthal tend to throw doubt upon the actual existence of the alleged similarity between human blood and that of the higher apes, and this circumstance renders untenable all the conclusions based upon this similarity, viz. that man is very closely related to the higher apes, or is even an ape himself. I should like to refer to some recent microscopical investigations made by Raehlmann into the red corpuscles. Those occurring in human blood present certain peculiarities that are not found in the blood of other vertebrates. In the course of his researches into the causes of sleeping-sickness, Brumpt found that, as a rule, the disease followed when other mammals were inoculated with the blood of sufferers from sleeping-sickness; a few apes and pigs alone were exempt.[1]

This is a very remarkable fact. Are we to infer from it that the composition of human blood differs most completely from that of the blood of some apes and pigs? Such an inference would manifestly be false; but this fact shows us how carefully we ought to proceed in drawing conclusions

[1] Cf. *Modern Biology*, p. 469. Notes of further research will be found in *La Nature*, November 17, 1906, pp. 390-392.

of this kind. In spite of my deep respect for the very ingenious research work that has been done on the subject of blood-reactions, I believe that in time scientists will cease to lay too much stress upon their phylogenetic importance.

Comparative morphology furnishes other reasons for regarding it as improbable that man is immediately related to the higher apes. Virchow, Ranke, Kollmann, and others pointed out some time ago that man and the higher apes seem to be the two extremes of widely divergent series of evolution. Man shows an inferior development of the hind limbs, whereas the ape has attained to a higher development in this respect, so that man cannot be closely related with the ape; both are representatives of divergent lines of development, and their common starting-point must be sought further back. If we try to trace the descent of man from apes, not necessarily from apes of any existing species, but from those of some extinct kind, we arrive at a real contradiction. Let us assume that the biogenetic principle is true, in the sense that the growth of the individual represents a faithful reproduction of the evolution of the race. Among the higher apes the young resemble man in the formation of the cranium, and in the shape of the face, far more closely than the old apes do, for in them the pithecoid characteristics are far more prominent. This fact, interpreted in accordance with the biogenetic principle, would lead to this conclusion: In their youth the higher apes pass through a stage of marked resemblance to man, hence apes are descended from man, not man from apes. The absurdity of this argument is manifest.[1]

I now proceed to the second theory, viz. that there is *some more remote connection* between men and apes. This is the theory put forward by Klaatsch, Stratz, Alsberg, and other anthropologists, who assume that a common ancestor lived in the old tertiary or pre-tertiary age, and was the progenitor of one line of descendants, who evolved into men, and also of another line, who evolved into the apes of the present day. This theory agrees with the facts of comparative morphology, with the different development of the extremities in man and in apes, and it is more probable from the zoological point of view than the theory of relationship by direct descent.

Are we therefore to accept the theory of indirect relationship without further question?

I think we ought to exercise great caution, for this theory is not as yet by any means proved. The hypothetical primitive form, upon which it is based, is very obscure. Klaatsch speaks of a 'general pithecoid type,' which gives rise to considerable difficulty, as it does not agree with the human type which is supposed to have originated

[1] Or have we here perhaps one of those famous 'counterfeit evolutions,' which, according to Haeckel, have been permitted by nature in order to falsify palingenesis by cænogenesis?

in it. Stratz presents us with a curious creature called a 'Molchmaus' (salamander-mouse) as the primitive type. In fact, so little seems to be known of the form of the common ancestor, that in 1899, at the Anthropological Congress at Lindau, when Klaatsch expounded his theory in detail, Ranke replied that such hypotheses were purely matters of imagination.

In criticising these theories, we must not lose sight of a very important *palæontological* consideration. The further back we place the ancestral form, *the more connecting links* must we assume between man, on the one hand, and this original ancestor on the other. Let us consider the two hypothetical pedigrees. We find in the one case the successive lines of evolution leading up to the apes and prosimiæ of the present day—a beautiful palæontological genealogy—thirty species of fossil prosimiæ and eighteen species of fossil apes. But if we look at the other line, where we should expect to find the intermediate forms between the ancestral type and man of the present day, we discover nothing. *There is not a single genus or species that can be regarded as a connecting link.* This fact is very important. If such a development ever really took place, we should surely find transitional forms also on this side.

Possibly some one will remind me of the *Pithecanthropus erectus*! I am just coming to it. In the course of *the palæontological examination of the human race*, it has happened again and again that connecting links between man and apes have apparently been discovered, but each time the discovery has led to disappointment. The *Pithecanthropus erectus* made a great stir. A Dutch army surgeon, named Eugen Dubois, in 1891 discovered in an old river-bed in Java, at a distance of several yards from one another, the vault of a cranium, a femur, and first one and then another molar tooth. He believed that these had all belonged to one and the same individual, who was neither a man nor an ape, but *something between the two.*

The lecturer displayed a diagram of the reconstructed cranium of the Pithecanthropus.

The discovery aroused much surprise. I remember how at the Zoological Congress of 1895, in Leiden, Eugen Dubois spoke for two hours, trying to prove that this Pithecanthropus was the hitherto vainly sought missing link between man and ape. During the proceedings Virchow, as honorary president, sat with a fixed, judicial expression. I tried to make out what he thought about the paper, but it was impossible. When Dubois had finished, he stood up, and in the most courteous terms expressed his grateful appreciation of the speaker, but added, that it would not be possible to arrive at any definite conclusion, until a complete skeleton was found. This somewhat severe condition still remains unfulfilled, and the hopes, to which the discovery had given rise, were disappointed.

For the amusement of his audience the lecturer here displayed a lantern slide on which was a grotesque figure representing the Pithecanthropus as a dandy, or masher. It was copied from the bill of fare at the third Zoological Congress.

The highest scientific authorities of the present time have decided, with regard to the Pithecanthropus, that he was *a true ape*, belonging to the Hylobatidæ group, which in many respects resembles man more closely than do certain anthropomorphic apes, although in other ways it is more closely allied to the lower apes.

Still more famous than the Pithecanthropus was the so-called *Neandertal* man, discovered in 1850 or thereabouts. His cranium was found in front of a cave in the Düssel valley near the Rhine. It was examined repeatedly, and anthropologists expressed a great variety of opinions regarding it, some taking it to be part of the skull of a Mongolian Cossack. Virchow even then expressed his doubts as to its really possessing the antiquity ascribed to it. In 1901 Schwalbe examined it again, and thought that he found support for the hypothesis that the owner of the cranium had not been a human being, but had belonged to some species standing midway between ape and man. Not long after, however, in 1904, the same scientist declared that the Neandertal creature had not belonged to any intermediate species, but was a man of some prehistoric race, resembling lower animals. He named this being the *Homo primigenius.* Now it is exceedingly interesting to trace the steps by which this homo primigenius, who was supposed to belong to some distinct species, proved finally to be only a member of some ancient race of men.

The lecturer here displayed a lantern slide showing the cranium of the Neandertal man (according to Schaafhausen), and then, referring to Macnamara's cranial curves on another diagram, he pointed out that, apart from size, there is no essential difference between the cranium of a chimpanzee and that of the Pithecanthropus. A comparison between the capacity of the crania of the Neandertal man and of an Australian negro respectively proved the difference to be extremely slight, less than that existing between the crania of an Australian black and of a modern Englishman.[1]

It is certain that there are great difficulties in the way of basing any argument in favour of the descent of man from beasts upon the crania that have been discovered.

The subsequent history of the Neandertal cranium belongs to the year 1905, when some very interesting discoveries of skeletons were made at Krapina in Croatia. Kramberger examined these remains very exactly, and compared the crania with the largest collection of fossil and recent crania ever placed at the disposal of a scientist, and he arrived at the following conclusion, which was published in the last number of the *Biologisches Zentralblatt* for 1905. Lest you should fancy that I represent the matter subjectively, I will give you some exact quotations. Kramberger sums up his arguments in the following words: 'There were

[1] See *Modern Biology*, pp. 476 and 480.

numerous stages of gradual transition between the homo primigenius and modern man, and yet the latter is connected with the former in such a way that an uninterrupted course of evolution extends from the homo primigenius to the homo sapiens, and this series of stages of evolution is of such a kind that (1) every characteristic, differentiating the homo primigenius from the homo sapiens, may occur in isolated cases also in human beings of our own day, and (2) conversely, the characteristics of modern man may occasionally occur in crania of the homo primigenius.' Kramberger deduces from this a *continuity of evolution* between the homo primigenius and modern man. I believe, however, that we have no mere continuity of evolution, but that the homo primigenius simply belonged to an older stock of the present human race. And why? Because there are *numerous stages of gradual transition* between the primeval man and man as he now exists.

Let us imagine for the sake of argument that some palæontologist discovered the skeleton of a fossil variety of dog, differing both from the still existing varieties and from fossils. He infers that he has proof of the existence of a new species of Canis, but subsequent investigations prove conclusively that all the peculiarities of the alleged new fossil variety occur in other fossils or in living varieties. Every zoologist would say at once that in the exact sense this was not a new genus, but only a *variety.* We must apply the same argument to the human skeletons. The zoologist must acknowledge that all the known forms constitute only one genus, and that is the genus of *Homo sapiens.* Homo primigenius must be included in it, and he must be described as *Homo sapiens primigenius.* He is the old diluvial human race.[1]

Next in order comes the *Homo sapiens fossilis,* and then man as now existing. All the evidence in support of the descent of man from beasts, which people have tried to deduce from the Neandertal man and his contemporaries, falls to the ground, as the primeval man appears to be *a true man* in respect of his body and of his mind.[2]

I appreciate fully the zeal with which scientists are carrying on their investigations into the primitive history of the human race; and provided they do so in accordance with scientific procedure, I have no reason at all for protesting. Whatever science reveals I shall accept without reservation, but the case is entirely different with phantoms of the imagination set forth as facts. Serious scientists, however, do not present us with such fictions, and in support of this statement I may refer to Professor

[1] According to C. Toldt, the absence of a true chin is a constant mark of this primeval race. On the subject of the formation of the chin, see the *Correspondenzblatt* of the German Association for the Study of Anthropology, Ethnology, and History of the Primitive Ages, 1906, No. 2, pp. 9-17.

[2] On this subject cf. Dr. Hugo Obermaier, *The Oldest Remains of the Human Body considered from the Point of View of Comparative Anatomy and Anthropology: Primeval Man on his Intellectual Side.* Vienna, 1905.

Branco, who delivered a remarkably fine and instructive lecture upon fossil man, at the fifth International Zoological Congress at Berlin in 1901. The chief points in this lecture are summed up in the assertion that *we know absolutely no ancestors of the human race*, for all fossil remains of human beings are the remains of genuine *men*, such as we are now. Branco at that time regarded the Neandertal cranium and that of Spy as exceptions, for such was the general opinion in 1901, but now it is certain that these prehistoric crania belonged, not to any ancestors of the human race, but to an earlier variety of the human race. Therefore Branco's statement: 'We know no ancestors of man,' is now still more true than it was in 1901.

Allow me to read you a quotation from Professor Schwalbe on the subject of primitive man. He is well known amongst modern anthropologists as one of the chief supporters of the theory that man is descended from beasts, but he is a thorough scientist. In the introduction to a work on the primitive history of man (1904), in which he upholds the descent of man from beasts, and connects the Dryopithecus, etc., with the ancestors of man, he says: 'In no department of natural science has the attempt to draw general conclusions from an aggregate of facts been so much influenced by the subjective opinions of the individual scientist as in the primitive history of mankind. On this subject it has frequently happened that views, based on a few facts, have been regarded as definitely obtained scientific results by those who have not studied the matter closely, because these views have been enunciated with a peculiar assurance.'

I fancy that Professor Schwalbe, if he were present, would not be offended if I said: 'The fate of Professor Schwalbe's homo primigenius has given fresh confirmation to this statement.'

I recognise the value of scientific research, and do not reject it in any hostile spirit, but I cannot say as much for the attempt to represent the descent of man from beasts as a conclusively proved fact, as Haeckel has often done, on the last occasion in the course of his Berlin lectures in 1905.[1]

I regret to have to say this, yet I am not opposing Haeckel personally, but his assertions regarding the descent of man, and my conscience compels me to continue this opposition.

I do not intend to discuss the matter further, and will content myself with reading to you one passage from his work entitled: *The Struggle regarding Evolution*, which contains his Berlin lectures. On p. 99 will be found the genealogy of the primates, in which is a perpendicular central line containing the direct ancestors of man. Their order is as follows:—As man's most remote ancestor Haeckel

[1] See his work entitled *Ueber unsere gegenwärtige Kenntnis vom Ursprung des Menschen* (Our present Knowledge of the Origin of Man), 1899, p. 22, and his *Welträtsel* (Riddles of the Universe), p. 97. Nevertheless, during the evening discussion, Haeckel's assistant, Schmidt-Jena, attempted to maintain that Haeckel's genealogies were merely stated as hypotheses. See Part II. of the present work.

gives the *Archiprimas*, a purely imaginary form. From him are descended the *Pachylemures*, represented by the universally accepted *Lemuravida*. Next come the *Necrolemures*, of whom nothing very definite is said. From them, according to Haeckel, are descended the apes, one of whom in particular, the *Archipithecus*, is the ancestor of man. (This is again a product of the imagination.) The *Prothylobates* is a direct descendant of the Archipithecus, but he too never had any existence, and has been invented; from him is descended the *Pithecanthropus alalus*, the speechless primitive man. He could not be called *Pithecanthropus erectus*, because scientists had already excluded this from the list of man's direct ancestors, so Haeckel calls him the speechless primitive man; but he too is an imaginary being. Next to him we have the *Homo stupidus*, the stupid man, who is a very real creature (laughter), and from him at last we arrive at the *Homo sapiens*. Only the last two out of the whole list are really known to exist, but I think we ought not to regard the *Homo stupidus* as the ancestor of the *Homo sapiens*. Comment is needless!

I have just shown *how great a difference there is between scientists who seek scientific proofs* and those who publicly proclaim the descent of man from brutes as *an absolute fact, basing such a descent upon an imaginary genealogy. The difference is indeed great*![1]

[1] Reinke, the well-known biologist at Kiel, remarks on this subject:

I may be permitted at the close of these three lectures to refer once more to the Christian standpoint, as I did at the end of my first lecture, and to set before you in conclusion what I may call a mental lantern slide.

Before me I see a huge ocean, and in its midst a towering rock, at the foot of which the waves dash up and retreat in endless alternation. This rock is the Christian theory of life, and the waves at its foot are the changing systems of human knowledge. The rock has stood firm and unshaken for thousands of years, whilst around it many a mighty storm has raged and died away. One such storm began 350 years ago. A wave had rested calmly at the foot of the rock for so long that the inhabitants believed it to be inseparably connected with the foundations of their rocky dwelling, and thought that the rock would inevitably be swallowed up in the deep, if another wave should come and displace the former one. At last a new and mighty wave did come, and displaced the other, but the rock stood firm. I think you will have no difficulty in understanding my picture. The storm to which I refer, when the wave of human knowledge raged against

'We merely have dust thrown in our eyes when we read in a widely-circulated book by Ernst Haeckel (he is referring to the *Riddles of the Universe*) the following words: "That man is immediately descended from apes, and more remotely from a long line of lower vertebrates, remains established as an undoubted historical fact, fraught with important consequences." It is absurd to speak of anything as a fact when experience lends no support to it.' (Haeckel's *Monism and its Supporters*, Leipzig 1907, p. 6.)

the rock of the Christian theory of life, was the struggle between the Copernican and the Ptolemaic systems. The Ptolemaic system had rested for centuries so peaceably beside the rock of Christianity, that men believed them to be inseparably connected, —that the rock must fall if the earth were to begin to revolve round the sun, and ceased to stand still. But the rock still stood firm when the old wave had to give way, and the new Copernican system, as the more powerful, expelled the Ptolemaic, and the earth really began to revolve round the sun. The minds of the faithful ceased to fear, for they saw that there was no reason for alarm ; the rock stood far too firm to be shaken by any transient surging of the waves.

Another three hundred years passed, and a fresh storm threatened the ancient rock. Once more had a wave rested in long-continued peace at the foot of the rock, and once more had those dwelling on it come to regard the wave as essential to their very existence, so essential that, should it give place to another and a more powerful wave, the downfall of the rock must necessarily follow. And the new wave came, and it will probably be victorious in the conflict now raging between it and the old. Will the rock fall, if the old wave is expelled ?

This picture, too, is easily understood.

I am referring to the theory of evolution as opposed to that of permanence, according to which

God created every variety of beast and plant in its complete and definite form. This old theory seemed so beautiful, and fitted so simply into the Christian theory of life, that many believed Christianity must inevitably fall too, if it were undermined.

In 1859 came the moment when a powerful wave, starting from England, assailed us like a deluge. It increased in strength and power until the foam flecked the very pinnacles of the rock. It is true that this wave no longer bears the name of Darwin and of the Darwinian system in the narrower sense, but it is the theory of evolution which is waging war upon the theory of permanence, and has hitherto been victorious in the strife, and will probably remain so to the end. Are we on this account to fear the downfall of the rock ? No—by no means. The rock of Christianity will stand firm, even if the theory of evolution as a scientific hypothesis triumphs over the old theory of permanence. Perhaps, after the lapse of some few decades, this new wave of science will come to rest peaceably at the foot of the old rock, and centuries later again another stronger wave will come and displace in its turn the theory of evolution, but the rock of Christianity will stand firm, as it has ever stood.

I for my part am convinced that nothing can shake the rock of Christianity, and my conviction rests on the fact that the waves and the rock are not natural enemies, but natural friends. *Human knowledge and the Christian faith are not opposed to one another Both are streams flowing from one original source, from one and the same infinite, eternal, and divine wisdom. This wisdom cannot contradict itself, although it may address us now in one and now in another language. Hence I am firmly convinced also that there can be no real contradiction between Christianity and science.*

PART II

EVENING DISCUSSION

Introductory Remarks on the Origin of this Discussion

Several newspapers expressed disapproval of the arrangements made for the discussion on February 18th. In order to remove the misapprehension existing on the subject, I propose to begin by giving a short account of the circumstances that led up to the discussion, although I need hardly say that I have no intention of hurting any one's feelings.[1]

I was induced to allow the discussion to take place because Professor Plate, a member of the German Monistic Association, refused otherwise to sanction the programme of my Berlin lectures. It was originally proposed merely to hold a discussion, but gradually the proposal developed into a plan for devoting a whole evening to it. On January 25th, 1907, the programme was published, and with the assent of the Committee the following general rules were laid down for the debate to be held on the evening of February 18th:—'Those who have previously given notice to the undersigned (*i.e.* the

[1] I gave a short account in the *Umschau* (1907, No. 17, p. 288) of what led to the evening discussion.

members of the Committee, see preface, p. viii), or to the lecturer, shall have an opportunity of stating and defending any opinions they may have at variance with those put forward by the lecturer. Some of the undersigned Committee members propose to avail themselves of this privilege.'

These were the only conditions officially laid down by the Committee; the details were left to be settled later on by the lecturer and his opponents.

The number of gentlemen who gave in their names as wishing to take part in the debate amounted to twenty-five, and amongst them were some who were not hostile to my views. It was obvious that so many could not speak on one evening, and I thought it best that preference should be given to my opponents.

In order to arrive at a definite agreement as to the conditions for the debate, on February 12th I had a conversation with my chief opponent, Professor Plate, who was a member of the Committee. I told him that I must be allowed to speak *at least twice*, once after him and again at the close, after the other speakers, and I stipulated that *the whole debate*, including my reply, should not much exceed *two hours* in duration.[1]

[1] I had originally declared that it would be only fair to allow me to give a few words of reply between the other speeches, if it seemed to me desirable, but eventually I withdrew this condition in order not to prolong the debate unduly, and I agreed to speak only *twice*, if the debate were *limited to two hours.*

After this conversation with Professor Plate, I saw another Committee member, Professor Waldeyer, the Privy Councillor, and communicated to him these conditions, professing my willingness to assent to them. Professor Waldeyer raised no objection, and promised to take the chair at the debate.

The conversation with Professor Plate, to which I have referred, was the only one which I had with my opponents as to the arrangements for the debate. I was therefore obliged to regard them as settled, and I had no reason for thinking that the majority of my opponents would alter them, without giving me due notice of their intention.

During the morning of February 18th, the day fixed for the debate, my opponents held a meeting, at which Professor Plate presided, as Professor Waldeyer was unable to be present. *I received no invitation to this meeting, and no intimation of it was sent me,* although the arrangements for the debate had to be made conjointly by me and my opponents. At this meeting the majority arranged *the order of the proceedings,* in spite of protests raised by a minority, which consequently expressed its determination to take no part in the debate. It was decided that I should be allowed to speak only *once,* and that at *the close* of the whole discussion, which, considering the length of time allotted to each speaker, was certain to last *for two hours and a half,* or even longer.

I was not informed of these new arrangements made by the 'majority,' until I arrived at the hall in the Zoological Gardens on the evening of the debate, when Professor Plate communicated them to me.[1]

I understood that the 'majority,' to which Plate alluded, was the majority of the Committee members, for in my opinion they, and they alone, were justified in making any arrangements, if I were to be excluded from all deliberations concerning them. Consequently I did not appeal to the President, Professor Waldeyer, against this alteration in the arrangements for the debate, although it was most disadvantageous to me. It was only on the following day that I learnt that the majority in question was the *majority of my opponents*, and that the organising Committee had not been consulted about the alteration, any more than I myself had been.

Professor Waldeyer took the chair at 8.30 P.M., and managed the debate with great discretion. About two thousand people were present, and followed the proceedings to the end with deep atten-

[1] Professor Plate wrote to me subsequently to assure me that on this occasion he had expressly mentioned the majority *of my opponents*. I cannot remember having heard these last words, but even if they escaped me, that circumstance does not in any way affect the subjective fact which I have just stated, viz. that I believed him to be speaking of the majority of the *Committee*. Professor Plate assured me further that, at the meeting of my opponents on February 18th, he had supported the arrangements to which on February 12th I had given my assent. I notice this assurance with gratitude.

tion, although there were many digressions from the subject proposed for discussion. Eleven speakers took part in the debate, and with the exception of Dr. Thesing, the last speaker, who did not commit himself, all were opposed to me.

I may remark that in compliance with Professor Plate's suggestion, just before the tenth speaker came forward, after it had struck eleven, the chairman suspended proceedings for five minutes, in order, as he said, to give any members of the audience who were tired an opportunity of withdrawing. Nevertheless the audience persisted in remaining to hear my closing address.

I need not say much here by way of introduction to my antagonists' speeches. Professor Plate spoke on the whole to the point, although his representation of my views was not free from obvious misinterpretations. To some extent, however, he strayed from his subject, and especially in his peroration he made a personal attack upon Wasmann as not being free to carry on his investigations.

Of the other speakers those whose remarks were most to the point were Professor Dahl, Dr. Juliusburger, Dr. Schmidt-Jena, and Dr. Thesing. Professor von Hansemann allowed himself to digress from the scientific subject proposed for discussion, whilst Count Paul von Hoensbroech's speech degenerated into a tirade against the Catholic Church. As his words had nothing to do with my subject, I may dismiss them without further consideration, and I took no notice of them in my closing address at the evening discussion.

Herr Itelson's historical platitudes regarding the 'crumbling away of the rock of the Church' call for no detailed comment. With these exceptions, my opponents' speeches are faithfully reported, and I have added critical remarks to them. My closing address was taken down in shorthand, so it can be given *in extenso*, with my supplementary remarks.

As my opponents had been speaking for three hours before I could begin my reply at 11.30 P.M., it was obviously necessary for me to express myself shortly, and to limit myself to about half an hour. This fact seems to justify me now in criticising my opponents' speeches, and my remarks here should be regarded as a continuation of the public discussion of my lectures.

I shall try to avoid all personal bitterness. Professor Waldeyer opened the debate at 8.30 P.M., and addressed the audience to this effect :—

LADIES AND GENTLEMEN,—

I have been asked to take the chair this evening during the discussion. I have agreed to do so, *assuming that the proceedings are to be strictly limited to calm, genuine, and scientific expressions of opinion*, and I beg that no speaker may be interrupted by any remarks. Such remarks could only cause disturbance, and would deprive our gathering of its dignified and scientific character.

I will begin by reading the names of the speakers, as they have been sent in to me.[1]

At the head of the list stands the name of Professor Plate, who will open the debate. He is a member of our Committee, and as he proposes to give a sketch of the whole subject under discussion, we have allowed him to speak for half an hour. Dr. Bölsche will next address you, and then Professor Dahl, after whom comes Dr. Friedenthal, to whom, in consideration of the subject which he has chosen, we have assigned twenty minutes. He will be followed by Professor von Hansemann and Count von Hoensbroech, to whom also we have assigned twenty minutes. Then come the names of Dr. Juliusburger, Dr. Plötz, and Dr. Schmidt-Jena. I do not know whether Professor Stumpf wishes to speak or not, but I shall give him the opportunity of doing so, as I was told that he intended to say something.[2]

Dr. Thesing will speak last. I have just been told that Mr. Itelson wishes to say a few words on the historical aspect of the matter. He will have his turn after Count von Hoensbroech.

All the speakers, with the exception of Professor Plate, Dr. Friedenthal, and Count von Hoensbroech, are limited to ten minutes, and I think we must adhere strictly to this arrangement, and therefore

[1] *i.e.* the names of those speakers who remained of the original twenty-five. See above, p. 86.

[2] When invited to speak at almost half-past eleven o'clock, Professor Stumpf declined to do so.

I earnestly ask you all to observe the time, and not to be offended if I hold my watch in my hand, and request each speaker to stop, as soon as his time has elapsed. Further, I ask you all to accept my ruling —otherwise it is impossible to control a debate of this kind.

I notice that a longer time is assigned to Father Wasmann, as he has to answer all the objections brought against him. His reply will close the proceedings. The list of speakers cannot be augmented, and each is allowed to speak once only.[1]

I will now declare the debate open, and call first upon Professor Plate to speak.[2]

I. Professor Plate's Speech.

I have often been in this hall, but I have never before seen such a sight here as I now behold. As a rule, entertainments are given here, or the good ladies of the West End assemble here to drink coffee, but, to-day, we are to consider the most serious questions. Father Wasmann of the Society of Jesus cast a spark into the Berlin world, which

[1] The *Vossische Zeitung*, in giving a report of the discussion, naïvely emphasises the fact that, of the general public, no one rose to speak on Wasmann's side.

[2] The speeches of my opponents are not reported word for word, with the exception of Dr. Juliusburger's, but only so as to show their drift, as far as it was essential to a due comprehension of the discussion.

The *Deutsche Tageszeitung* and other papers have given detailed accounts of the views expressed by my opponents.

has grown to a mighty flame. His action has provoked this meeting, and the enormous attendance proves how deeply interested people are in the subject under discussion. We are dealing with *the old conflict between the Church and Science*—the conflict which has raged for centuries, and to which men like Galileo, Columbus, etc., have more or less fallen victims. Father Wasmann's peculiar position gives the struggle an unusual character on this occasion. The priest, the representative of the Church, appears under the form of a scientist. Hitherto the Church has approached us only under the guise of the priest, now she consents to deal with natural science, and we scientists welcome this as a hopeful sign, pointing to the possibility of a reconciliation between us. It is true that such serious questions cannot be decided at large public meetings, but Father Wasmann wished to hear the critical opinions of his fellow-workers, and so I said to myself: 'We scientists ought not to refuse to state our convictions. It may be impossible to avoid hard words, but they are aimed not at the man, but at his subject.'

In my closing speech I dealt fully with the objections raised by this my chief opponent, and therefore, in order to avoid repetitions, I will limit myself here to the most indispensable remarks, and will content myself with pointing out that in my lectures *I did not touch upon the conflict between the Church and Science*, but sought merely to throw some light upon the question: 'What are we to think of the theory of evolution?' At the very beginning of his address, Professor Plate took up a wrong standpoint, and transferred what was originally a purely scientific argument to the field of religious controversy, thus rendering the matter more interesting. Even Columbus is dragged in as a victim to the conflict between Church and Science, although he always declared the propagation of the Catholic faith to be the chief reason for his voyages of discovery, and although he died wearing the Franciscan habit.

Professor Plate went on to discuss the impression made upon him by my lectures. 'Our lecturer,' he said, 'has a *twofold character*, he is *a remarkable combination of a scientist and a theologian*; both contend about the same object, but in the contest the theologian invariably prevails and the scientist succumbs. Father Wasmann must say, in the words of the poet, that "two souls dwell within him." He speaks of inheritance, of adaptation, of rudimentary organs, and in fact he avails himself of all the assistance afforded by zoology, as long as he is dealing with a topic on which the Church has not laid hands, and with regard to which she has uttered no final opinion. But no sooner does he touch upon such a subject as the origin of the human race, than he brings dogma to bear upon scientific research, and we disapprove of this.'

In ascribing to me a twofold character, Professor Plate was not original. Several of the monistic critics of my book *Modern Biology and the Theory of Evolution*, such as Forel, Haeckel, von Wagner, etc., have done the same, and have adopted the easy method of putting down all that displeased them in my views to my being a theologian, and all that pleased them to my being a scientist. It is obvious that in this way they can prove nothing against me. A man who is at the same time a scientist, a philosopher, and a theologian, must examine a complicated question, such as the theory of evolution, from these various points of view, and not merely from one; but in as far as certain knowledge is to be derived from these various aspects, they cannot contradict one another, because there is only *one* truth. Plate thinks that if Wasmann the scientist comes into conflict with Wasmann the theologian, the former always lays down his arms; but many a theologian might with equal right bring exactly the contrary charge against me, and assert that I am constantly striving to make my theological opinions agree with the ascertained facts of science. In making this remark, Plate seems to have completely forgotten that in my third lecture, in which I dealt with the origin of man, I had recourse to zoology in just the same way as in the first lecture, when I spoke of ants, beetles, etc. In none of my lectures have I used dogma against science; what Plate here calls dogma, is really the law of reasoned thought, which I have repeatedly used against the false conclusions of a monistic philosophy.

Professor Plate next proceeded to establish more firmly what he had said about my twofold character. He referred to *the problem of the existence of matter*, and said: 'We scientists maintain that matter exists, that nothing is formed out of nothing, and that matter is everlasting. We cannot accept the theory that matter was created, and if we did accept it, we should be no better off. We are modest enough to dispense with a further solution of this problem.'

Although Professor Plate professes to speak in the name of scientists in general, he does not do so, but he is expressing his opinion as a monistic philosopher. The scientist may say: 'I know nothing about the origin of matter,' but he must not say that matter is therefore everlasting; for a statement of this kind belongs to philosophy with its metaphysical problems. To assume the eternity of matter is wrong philosophically, because only an infinitely perfect Being—God—can have in Himself the reason of his existence. Therefore we must admit the creation of matter by God. (Cf. my second lecture.) The admission of creation does not furnish us with a *scientific* explanation (for this is not possible in the case of metaphysical problems), but it gives us a *philosophical* explanation. Plate's argument: 'Matter exists—nothing is formed out of nothing—therefore matter is everlasting'—is quite contrary to philosophy. Certainly it is impossible for anything spontaneously to proceed from nothing, but a finite being can begin to exist, if it is called into existence by an infinite being.

Professor Plate went on to say that I had discussed the *origin of living creatures*, and that two views were opposed to one another on the subject. 'We scientists maintain that *there must have been a beginning* of life, but that to assume creation would not be to account for it. We ask further whether we can penetrate more deeply into the subject. If we have *points d'appui*, we are justified in setting up an *hypothesis*, of the conditional truth of which we are convinced. We all admit that we are not yet able to observe the manner in which living beings have proceeded from inorganic matter, but we may lay down the hypothesis that in some previous age and under other conditions then prevailing on the earth, and perhaps still prevailing in its interior, living beings came into existence from inorganic matter. In the second place, we say that protoplasm consists of twelve elements, and albumen, which is characteristic of living substances, only of five. If at death the body can be resolved into dust, there must be certain conditions under which it can come into existence from dust. Thirdly, we maintain that there are transitions between organic and inorganic matter. There are substances which display properties otherwise possessed only by living beings. Thus crystals can grow and reproduce themselves. We even know of *liquid crystals* which move, subdivide, consume one another and unite with one another, just as living beings do. These are *ascertained facts* about which there can be no dispute, and they serve as a basis for our hypothesis that *living beings have at some time proceeded from inorganic matter*. Father Wasmann only propounds problems on this subject when he asserts that these things *were created.* Creation is to be a comfortable solution which ought to satisfy us as scientists, but we regard the origin of life as a *zoological problem*, in exactly the same way as we regard the origin of the Alps as a geological problem, and we refuse to be deprived of this problem.'

In answer to these remarks on the origin of life I will only say shortly :—

That there was a time when living beings came into existence is obviously true, and we may assume that they were formed from inorganic materials, not by way of direct creation. But the question is, whether they could of themselves have come into existence from inorganic matter spontaneously or not. Scientific facts disprove the possibility of spontaneous generation, and therefore, in the present state of our knowledge, we are justified in assuming that the first organisms came into existence in consequence of the Creator's influence upon original matter. The contrary hypothesis, that, viz., of *spontaneous generation*, cannot claim even scientific probability. This becomes plain when we examine the arguments which Plate brings forward to support it. Reinke showed long ago that it was absolutely unreasonable to assume that, in some previous age, completely different conditions, which would have rendered spontaneous generation possible, prevailed upon our earth.[1]

That such conditions perhaps still prevail in the interior of the earth is not to be imagined, as the intense pressure would inevitably destroy all organic life. In the second place, Plate suggests some definite elements as the living constituents of organisms. If an organism at death is resolved into these constituents, it

[1] Cf. Wasmann, *Modern Biology*, pp. 207, 208.

certainly follows that *it was composed of them, but it does not follow that it came into existence out of them spontaneously.* Otherwise a house, which is pulled down and becomes merely a heap of bricks, ought to have come into existence automatically out of a heap of bricks. Thirdly, Plate refers to the alleged *transitions* between organic and inorganic substances. *Solid crystals* show no such transitions, for they aim at stable equilibrium of their molecules, as opposed to the unstable equilibrium of the molecules in living beings. Lehmann's famous *liquid crystals* do not supply us with the desired proof. The resemblance between their movements and those of the lowest organisms is *purely superficial,* and rests merely on chemical and physical modifications. Our imagination may lead us to speak of these formations as devouring one another, as having sexual intercourse with one another, and so on, but we cannot use these expressions in their literal sense. I shall refer to this subject more in detail in my closing speech and in the remarks upon it.

What Plate calls 'ascertained facts' do not therefore furnish us with any satisfactory scientific basis for the hypothesis of spontaneous generation. When he describes the assumption of a creation as a 'comfortable solution,' which he as a scientist is unable to accept, he ought to remember that we have no scientific explanation of the origin of life, and must therefore have recourse to a philosophical explanation, if we are to have one at all. I cannot agree with Plate, when he says that the origin of life is as much a question of zoology, as the origin of the Alps is a question of geology, for until the first organisms came into existence there were no laws governing organic life, but there were only the chemical and physical laws, and so no zoological problems could exist. But before the origin of the Alps the forces and laws of geology were already in action, and so a geological problem is in this case really presented to us. The origin of the first organisms may be described as a chemical and physical problem, but not as zoological or botanical. Moreover, the highest scientific authorities—such, for instance, as Professor Branco, in the course of his entrance address as a member of the Berlin Scientific Society (*Reports of Proceedings*, 1900, pp. 679-690)—have frankly declared that from the scientific standpoint we know nothing at all about the first appearance of life.

Professor Plate went on to say: 'Another equivocation on Father Wasmann's part appears in his opinions regarding species. He occupies a peculiar position in this respect. He has studied certain kinds of animals very closely, and has arrived at the conviction that the doctrine of evolution is theoretically sound. I regard it as showing an immense advance that a Catholic priest should venture to come to this conclusion, and I regret that so many Protestant clergy are unable to do the same. But at this point the Church suddenly intervenes, and reminds Father Wasmann that the types were created, and therefore a compromise is necessary. So he adopts the theory that the Creator once for all created certain types, and that these have subsequently developed. It is plain that he fails to reconcile these opposed theories, the *theory of permanence* and *the theory of descent or evolution.* If we examine what Wasmann considers as *natural types* originally created —I should prefer to call them *supernatural*—they prove to vary incalculably among themselves; sometimes we have large groups, sometimes small. Sometimes we are told that God created a *primitive ammonite,* then a *primitive horse,* a *primitive ant,* and so on. I will not argue about metaphysics, but I challenge the attempt to answer a purely zoological question with metaphysical phrases of this kind.'

The speaker might have saved himself the trouble of thus stating my views on polyphyletic evolution, if he had studied more closely what I really think on the subject. He might with advantage have referred to the third edition of my book on *Biology and the Theory of Evolution*, pp. 303 *et seq.*, and he would there have found that the 'natural types' are identical with the evolution series or pedigrees of the theory of descent, and their number, extent, and form are the subject of further biological research. The conceptions therefore are quite natural and by no means 'supernatural.'

He might also have found without trouble that, when he ascribes to me the idea that God created the primitive ant, ammonite, and horse, he is really speaking of the products of his own imagination. Finally, he would then not have confounded, as he has done, *the theory of permanence with that of creation.* In the book to which I have referred, as well as in my first two lectures, I showed plainly enough to convince all who were willing to understand me, that the theories of creation and evolution are not incompatible, but they harmonise and complete each other. It would be a difficult matter to find in any work of mine any attempt to solve the problems of zoology by means of 'metaphysical phrases.' I may refer also to the passage in my closing address in which I alluded to this subject.

The speaker went on to ask how Father Wasmann

had arrived at this view, and he declared it to be *the dogma of the creation.* 'This is the more remarkable as this dogma is based upon the Bible, and Father Wasmann tells us that we must not look for scientific accuracy in the biblical account of the creation. I think it altogether wrong to attack the Bible in any way, and I fully agree with Father Wasmann in asserting the Bible not to be a text-book of natural science, nor a manual of zoology or astronomy. But what is the result ? We are not to refer a zoological question to the Bible ! We scientists have nothing whatever to do with the Bible. Father Wasmann has some scruples about admitting constant interference on the part of the Creator, for such interference would be *degrading to Him.* For my part, if a Creator exists, I fail to see why He should not always interfere. If we admit that the Creator works through natural laws, we need only assume that the laws were in the beginning laid down by the Creator.'

My answer to Professor Plate is as follows : Even if there were no Bible and no dogmas of the Catholic Church, as thoughtful scientists we should have to ask ourselves whether we are to assume a *monophyletic* or a *polyphyletic* evolution of organisms. The reasons determining my answer to this question are scientific. (See p. 15 in the first lecture.) In saying therefore that the biblical dogma of the creation forced me to accept polyphyletic evolution, Plate has made an unfounded assertion.

With regard to the 'zoological question' which I am supposed to have referred to the Bible, I may say that the contradiction between the Bible and science has been insisted upon by my monistic opponents, by men such as Haeckel and Dodel, and I was quite justified in resisting their attacks.

Finally, Plate believes that some 'degradation of the Creator' is implied if I do not admit that He is constantly interfering with the laws of nature. This scruple rests upon the speaker's ignorance of the Christian Theodicy, and I should like to explain it to him shortly. It is a mistake to think that God could not constantly interfere with the laws of nature, but as God is *absolute intelligence*, His power is co-extensive with His wisdom. *The laws of nature are the expression of His wise design as Creator*, and He does not arbitrarily interfere with them, as to do so would be to disturb the *natural* order which He willed to exist. It is only to accomplish some higher, supernatural design that God can will to set aside the laws of nature, thus effecting a *miracle.* For those who possess the requisite preliminary knowledge of theology, this statement suffices to show that a miracle is in strict conformity with reason.

The speaker next discussed the principle of beneficial design in nature. 'I am glad,' he said, that 'Father Wasmann accepts the principle of *selection.* It is notorious that the vitalists hold another opinion, and think that a designing principle is inherent in the organisms themselves. These two views are as irreconcilable as fire and water. We may ask whether an organism *invariably* acts for its own advantage, or whether its actions result disastrously. Unfortunately in innumerable cases its actions are disastrous, as soon as it is placed in extraordinary conditions. There is therefore no immanent directing principle in nature. You all know this from the hackneyed saying that the world is a vale of tears!'

> Professor Plate has given an inaccurate account of my opinions regarding the theory of selection. (Cf. p. 41, etc. in my second lecture.) I believe it to be merely a subordinate *auxiliary factor*, and it assumes, as chief factor, ability on the part of the organism to produce forms adapted to the purpose in view, for otherwise no 'selection of the fittest' would be possible. Thus the directing principle, immanent in the organisms, and the principle of selection are far from being as irreconcilable as fire and water, but are the complement one of the other. Plate has plainly a mistaken idea of the immanent principle of vitalism, which he seeks

to disprove. He represents it to himself as an *absolutely* beneficial design, working out its purpose without reference to any external conditions of evolution. In reality there is in the organisms only a *relatively* beneficial design, dependent upon definite external influences, and so possessing definite limitations. If an organism is brought into unusual, and for it *unnatural*, surroundings, it is obvious that it will not always act beneficially. Plate has not proved anything against the existence of an immanent directing principle in nature.

In order to show that merely selection and not design controls the organic world, Professor Plate had recourse to the following simile. 'If an engineer wishes to construct a pump, he follows the principle of natural selection if he puts together the parts of the pump without much consideration, and makes perhaps two hundred pumps in hopes that one or other may chance to answer his purpose. No one would say that a man of this kind acts *with design*, but this is how nature acts according to the principle of selection.'

I should like to contrast this picture of Professor Plate's with another, far better suited to illustrate the real relation between design and natural selection. Let us imagine that some company offers a prize for the pump that best answers a given purpose. Various

> engineers set to work, and each constructs a pump. The company proceeds to test the pumps, and selects *the one that best fulfils the required conditions.* Although only one engineer receives the prize for his pump, all have kept *the appointed end* in view throughout their work.

Professor Plate proceeded to say that Father Wasmann spoke next of the doctrine of evolution, and referred to the *monistic view of the cosmic position of man.* 'In this case,' he remarked, 'I do not consider Father Wasmann to have been inconsistent, but he has not sufficiently emphasised the contrasts, which are sharper than he states. They amount, in fact, to this: the monist *asserts nothing* about the nature of God, but limits himself to the laws of nature. These laws are, indeed, the only things that we can establish with certainty; with regard to what underlies them there are many different opinions, and we monists are not all agreed on the subject. *Personally, I always maintain that, if there are laws of nature, it is only logical to admit that there is a lawgiver.* But of this lawgiver we can give no account, and any attempt to give one would lead us into unfounded speculations. It is there that faith begins, and many of us have given up all faith. For my part, I do not feel compelled to do so, but we must allow each man to act as he thinks right.'

I regard this statement regarding monism as the most important part of Plate's whole speech, and for that reason I append to it the following critical remarks:—

The opinion which Plate describes as 'monistic,' belongs, strictly speaking, not to monism, but to agnosticism, for the latter limits itself to the investigation of the laws of nature, without committing itself to any statement about God, who seems to the agnostic incapable of being known at all. Monism, on the contrary, asserts the absolute identity of God with the world, and thus professes to know something about God, although it is something wrong. Professor Plate is right in pointing out the great confusion existing in the monistic views of God, which are all at variance.

Plate's own confession that *where there are natural laws, there must be a lawgiver*, is of the utmost importance. A lawgiver *underlying* the laws which He has made, cannot be identified with those laws, for otherwise He would be superfluous, as the laws of nature would suffice *independently* of Him. Therefore the originator of the laws of nature must be *an exalted and intelligent being*, in fact, *the personal Creator* recognised by theism.

In my closing speech I laid great stress upon the fact that, by making this important concession, Professor Plate, a member of the

German Monistic Association, *has acknowledged himself to be a theist.*

It is true that he at once went on to add that we could assert nothing about this lawgiver without falling into unfounded speculations. But when a man has once grasped the fundamental element in the theistic conception of God, that *behind* the laws of nature is a lawgiver, not capable of identification with these laws, he is forced to think of God as an *intelligent and personal being,* unless indeed he admits his inability to think logically. *Faith,* however, does not begin at this point, as Plate imagines, for this is only the *foundation* of faith, which is concerned with supernatural revelation and not with natural knowledge. But whoever has once recognised God as the lawgiver of nature, cannot avoid asking himself whether this lawgiver may not have imposed upon reasonable beings *other laws* besides those of nature. To ask this question is not merely a matter of necessity as Professor Plate assumes.

Professor Plate next compares the *monistic* and the *theistic* views of the laws of nature. The *monist* says that there are only natural laws, and we do not know what underlies them. They are everlasting and *inviolable.* In this sense there are no *miracles,* there can be no violation of these laws;

we cannot assume that there is any arbitrary and capricious interference with the orderly course of events in the universe. The *theist*, on the contrary, says that there is a personal Creator, who imposed the laws of nature and therefore can alter them arbitrarily at any moment. This is perfectly logical. Belief in miracles cannot be uprooted by logic. But we may question nature, and ask ourselves whether we can really observe any instance in which the laws of nature are set aside.

As evidence against the occurrence of miracles, the professor related an experience of his younger days, which made a great impression upon him. He was standing once in Rome on a bridge over the Tiber, and was talking to a priest about miracles. The priest pointed to the river, saying: 'Look, the Tiber flows as a rule downward, but just at that spot it makes a counter-current.' Professor Plate answered: 'My worthy friend, if that is a miracle, then certainly miracles do take place.'

I believe that the question of the possibility and actual occurrence of miracles cannot be settled so simply as Professor Plate here suggests. In the first place, we must notice that the monist who proclaimed the *eternity* and the *inviolability* of the laws of nature, is asserting far more than he can prove. The utmost that he is justified in saying is that, as a scientist, he knows of no beginning to these laws and no exceptions to their application. But no sooner does the scientist confess, as Professor Plate has done, that a lawgiver underlies the laws, than he is forced to acknowledge the *possibility* of miracles, inasmuch as this lawgiver, having some higher supernatural object in view, may will to set aside the laws of nature under exceptional circumstances.[1] Plate is quite right in saying that belief in miracles cannot be uprooted by logic, especially as a miracle is no arbitrary and capricious interference with the orderly course of events in nature. That miracles are not of frequent occurrence is obvious, therefore Plate can scarcely require us to have daily opportunities of observing them; but one who seriously is in search of truth must ask himself whether *history* does not supply us with some undoubted instances of setting aside the laws of nature. The resurrection of Christ, which is the historical foundation of all Christianity, is a miracle of this kind.

In the last section of his speech, Plate attacked the '*rock of the Christian Theory of the Universe.*' He says: 'Father Wasmann calls the Church the rock round which the waves surge. I feel bound to contradict him. Think of the first wave in the time of

[1] My third opponent at the evening discussion, Professor Fr. Dahl, upheld the possibility of miracles in his work on *The Necessity of Religion*, p. 107, 1886.

Copernicus; it made an enormous breach in the rock of the Church, for the authority of the Bible was shaken for the first time. The new learning showed that the Bible contained errors, and could not therefore be the outcome of divine revelation. The Reformation, the second wave, made another breach, and now we have the doctrine of evolution, and this wave has destroyed belief in miracles. But does this imply the ruin of Christianity as a whole? No, only its purification or enlightenment; and I hope that science will lead to such an evolution of both Protestantism and Catholicism, that they will eventually unite and form one universal church.'

Professor Plate's whole wording here is rhetorical rather than logical. He is mistaken in thinking that a breach was made in the rock of the Church by the Copernican system. Copernicus did not prove the Bible to be wrong, but only showed that certain passages in it had to be interpreted in a way differing from the hitherto usually accepted manner. If this were not the case, Professor Plate would be wrong if, at the present day, he still speaks of sunrise or sunset. The language of the Bible is that of ordinary mortals. It is difficult to see why Professor Plate attacks the Bible here, when just before he remarked that he thought it altogether wrong to attack the Bible in any way (p. 104), and that he fully agreed with me in asserting the Bible not to be a text-book of natural science, nor a manual of zoology or astronomy.

Plate's second wave, the *Reformation*, has plainly nothing to do with the discussion of my lectures, as it was in no sense scientific. That the third wave, the *doctrine of evolution*, treated as a scientific theory, is not opposed to the Christian theory of the universe, I have proved conclusively in my lectures. Only the frothy dogmatism of monism is really inimical to Christianity, and it has nothing in common with the scientific doctrine of evolution.

Plate's expressed desire for an end to the sad religious differences in Germany is admirable, and I cordially agree with it (cf. the remarks on the subject in my closing address), but I fail to see what part *natural science* is destined to play in the accomplishment of this glorious task. Professor Plate's words seem to me obscure. Does he mean that we are to continue to demonstrate to mankind that there is no real antagonism between science and Christianity, as I attempted to do in my Berlin lectures? If this is his meaning, his opposition to my attempt at conciliation is inexplicable. Or does he mean that the monists ought to be in future less pugnacious, and cease to use science as a kind of battering-ram against

Christianity? If so, he has not set a good example in his own speech.

Or, lastly, does he mean that mankind is, in course of time, to be so far intellectually developed by means of the natural sciences, that men will thenceforth think only scientifically and lose all craving for transcendental ideals? If this were the case, science might claim to have triumphantly succeeded in uniting peaceably into one single, great, and universal church, free from all creeds, not merely Catholicism and Protestantism, but also Judaism and Mahommedanism, Brahminism and Buddhism, Confucianism and Taoism, Monotheism and Polytheism, Deism and Pantheism, Fetishism and Atheism. I fear, however, that this universal religion in its spiritual aspect will prove to be nothing but Atavism, *i.e.* a relapse into an animal form of religion, although the new German Monistic Association may profess the warmest admiration for it. The idea is, moreover, not new; it occurs in an extremely old Chinese legend.[1]

Professor Plate ended his half-hour's oration with these words: 'I repeat what I said before. Father Wasmann has a twofold character. He is at once a scientist and a theologian. He speaks as a scientist when he is discussing his own special

[1] Cf. A. H. Smith, *Chinese Characteristics*, p. 377. Shanghai, 1890.

department of research, viz. the inquilines of the ants and termites; but as soon as he touches on problems which the Church claims to have solved, he suddenly adopts a totally different method. The ground for this difference is undoubtedly his *voluntary or involuntary dependence upon the Church; to him is lacking the first condition essential to a true scientist, viz. freedom to think and to draw conclusions.* Although I gratefully recognise the fact that Father Wasmann, a Catholic priest, accepts in principle the doctrine of evolution, I am constrained to declare him to be *no true student of nature and no genuine scientist.*'

In order to prove the truth of his closing statement, the speaker ought at least to have shown that *I had arrived at scientifically false results, in consequence* of the dogmatic servitude with which he reproaches me. In the whole course of his speech he has not succeeded in doing this. It is probable that many members of his scientific audience, on hearing his last words, felt inclined to ask why he allowed his name to stand on the programme of Father Wasmann's lectures, if he did not regard the latter as a true student of nature and a genuine scientist.

A Protestant reporter, Dr. M. Senff, in the *Harzer Kurier* of April 27th and 28th, criticised Plate's speech rather sharply. He remarks that it contains 'a touch of something not quite straightforward.' (Cf. the extract from his critique in the supplement to this work.) Professor Plate seems not to have been aware that throughout his speech, but especially at the end, which was by no means to the point, he spoke *as an adherent of monism, either voluntarily or involuntarily.* For this reason we cannot regard him as free from prejudice or partisan feeling in the matter.

II. Dr. Bölsche's Speech.

Dr. Bölsche began by saying that, owing to the short time assigned to him, he proposed, in answer to Father Wasmann, merely to state *what he personally believed* to be the truth regarding the facts collectively. He continued: 'Father Wasmann has to a certain extent accepted the doctrine of evolution as applicable to the organic world; but, with regard to its application to man, he nevertheless insists upon the fact that we have as yet no satisfactory evidence of the descent of man from beasts. Then with a *salto mortale* he arrives at this assertion: "See, ladies and gentlemen, Christianity will stand firm as a rock towering above all these waves." This was the point which Dr. Bölsche wished to discuss. He said that if any one had advanced so far as to acknowledge the possibility of a natural evolution, he would be carried away *by quite definite logical sequences of thought*, and could find no obstacle in the fact that some little bones were missing in man, which would prove his descent from beasts. *Logic* has to continue its task unaided, and logic is the power that moves mankind. If it be once granted that, with regard to his body, man may possibly be descended from beasts, we are at once involved in the difficult question of the connection between body and soul. Whether we regard the world merely as a special soul, or whether we regard it as a panpsychism, in which we men are something objective, though at the same time inwardly intelligent,—*body and spirit are always in union*. This is *the first logical result* at which we arrive.'

The following criticism on this 'first logical result' must suffice. The question as to how far we must regard the theory of descent as proved, is not a matter of possibilities, but of facts. If, as Bölsche admits, natural science does not give us any actual proof of the descent of man from beasts, so far from being logical, we should be most illogical, should we, as scientists, assume this descent to be a fact. We should be still more illogical if, in consequence of the close connection that exists between the human body and soul, we were to conclude that man in his *spiritual* nature is the descendant of beasts, because we have assumed him to be such in his bodily nature. It is a perfectly illogical assumption on the part of monism to regard the body and soul as being merely two aspects of one and the same reality.

Bölsche's *second argument* is that we must penetrate deeply into the *animal kingdom*. 'If we consider the mind and soul of an animal, we find it impossible to distinguish clearly between the soul of an animal and that of a man. 'In the former exist the germs of all that makes for good or evil in the latter. They rise from the lower to the higher, often in so unworthy and mean a manner that we men with our intelligence must feel ashamed at being descended from the intelligence of a beast. Even Darwin declared that he would prefer to be descended from a cur that defended its master, than from a man who ill-treated his wife and children and killed his enemies. In his last lecture Father Wasmann spoke of the fragments of bone from Krapina, which point to some prehistoric cannibal feast, and according to him we must be the descendants of such cannibals. I am of opinion that there are many phenomena in the souls of beasts far higher and greater than the meanness that we detect in ourselves. *The most miserable thing which we see in the whole world is a degraded and debased human soul*, and the soul of a beast rises far superior to it, and reveals to us *maternal love* in its purest form ; so that the idea becomes an ennobling one that we, with our cultured human souls, have *ascended* from the souls of beasts.'

It is not easy to follow Dr. Bölsche's second argument, for it is altogether based upon that uncritical system of judging beasts by the standard of mankind, which Wilhelm Wundt designates as 'Popular Psychology.' That the physical impulses of animals occur also in men, is a well-known fact, but it is a mistake to assume that a beast possesses human reason and free-will. The man who, instead of using his higher intellectual capacities, abandons himself to his bestial impulses, sinks, it is true, *below the level of the beasts*, and his soul is indeed *more wretched* than that of an animal. In making this statement Dr. Bölsche has himself expressed a scathing condemnation of the opinion which he put forward in his work entitled *Das Liebesleben in der Natur*, regarding the ethics of mankind. The morality of beasts, if transferred to human beings, must inevitably lead to their brutalisation, and there can be no further reference to *Ideals*.

Bölsche's assertion that the people in Krapina were cannibals, is not accurate ; Dr. Hugo Obermaier examined the remains of bones and showed that it was a mistake to suppose that they were split open lengthwise in order that the marrow might be exposed.[1]

Let us now follow Bölsche's logical arguments as they advance in their triumphal career.

'If man has once arrived at the conviction that he with his intellectual soul has ascended from the souls of beasts, he grasps also the magnificent idea that everything in the world is determined by *natural laws. Universal logic* is based upon the universality of these natural laws, and everything depends absolutely upon this logic. If the smallest particle of it is abstracted from the universe, *everything* falls to ruin, not only the firmament of heaven, not only matter with its wild movements, but even our ideals—the best things that we possess—are shattered, if we take away logic from them. If this ' interior logic ' has once taken possession of a man, it will urge him to enthusiasm for *research*, in contradistinction to revelation. We need no other revelation than research, and in research into nature lies *the divine power*. ' Where research work is carried on, God is present, and that is the real sanctuary of mankind ' (Vischer). Whether we call this sanctuary *nature* or *God* is quite unimportant—the one word is spelt with six letters, the other with three, that is the only difference.

[1] 'La Station Paléolithique de Krapina': *L'Anthropologie*, xvi. p. 13, 1905.

And if this logic completes its work in the human soul, what will be the result? Will *Christianity* still stand firm as the rock against which the waves surge? That depends altogether upon what you understand by Christianity. All fixed dogmas will certainly fall—research will sweep them away by means of its logic—and the theory of life, which will stand firm, will not be the Catholic, nor the Protestant, nor the Jewish, nor the Mahommedan, but *the theory of the ideal*!' 'If you choose to call this ideal Christian, very well, you are entitled to do so.' In conclusion we have a little Bible lesson. 'Read the New Testament. Two things are necessary for man, love of God and love of his neighbour. If he possesses these, he possesses the highest to which he can aspire; on the one hand, he is in union with the universal principle which we may call God or nature, just as we please; *universal harmony* to its utmost extent—*this is love of God*; and on the other hand, he possesses *love of his neighbour*, the ideal humanity, which begins among the brutes and ends with man in his social and intellectual aspect. If you like to call this *the Christian theory* of the universe—well and good! This will be the rock against which no wave will dash, or rather, it will not be the rock in the midst of the waves, but it will be *the highest wave of evolution*, which certainly cannot be overtopped by any lower wave!'

These remarks seem not to require a long

criticism. We must remember that the speaker professed to give us only *his own confession of faith.* In spite of his fine words, such as 'universal logic' and 'universal harmony,' no one will discover in his speech any consistent logical arguments, and still less any really ideal theory of life. What Bölsche has given us is *no scientific result, but a dogmatic outline of monism,* in which all clear conceptions and ideals vanish in fantastic vagueness. The parody of the Christian commandment to love God and one's neighbour, with which the speaker concluded, was probably a severe trial of patience to the Christians among his audience.

III. Professor Dahl's Speech.

Professor Dahl began by saying that he intended to discuss only a few points, which seemed to him of particular importance, and with regard to which he did *not* agree with Father Wasmann.[1]

'In the first place, Father Wasmann has declared it to be incompatible with scientific thought to assume that *matter always existed.* I believe that

[1] Dahl seemed to imply that there were other points on which he agreed with me.

we have here some confusion of ideas, and that Father Wasmann probably intended to pronounce this assumption incompatible with scientific *concepts*. We can think of many things of which we can form no definite concept—billions and millions, for example. Thus we cannot imagine what is eternal or infinite, but we are quite able to think of it. In the same way as we are unable to form a concept of the infinity of space or of the indestructibility of matter, we are unable to imagine scientifically any origin of matter out of nothing. We cannot advance on these lines.'

Professor Dahl's first objection rests upon a misunderstanding on his part, as a reference to my second lecture (pp. 27-33) will show. I rejected the theory that matter always existed as being incompatible with scientific *thought*; I made no reference to scientific *imagination*, for it is obvious that we can only think of what is eternal and infinite, and cannot imagine it. On this subject see my closing address.

Professor Dahl went on to mention some real divergencies between my views and his own.

'Father Wasmann believes that *God intervened three times* in the course of the Creation. He first created matter, then He created life upon the earth, and lastly He created man. I will refer to man first. Father Wasmann tells us that we *cannot prove* man to be connected with beasts. This is a correct assertion; for the present at any rate we cannot speak of any definite proof—and we must proceed in another way. Let us say: there are two possibilities—*Either* man is genetically connected with the animal world, *or* he is not; there is no third alternative. We have to decide in favour of one or other of these possibilities, and, I ask you, which of them is *the more probable* from the scientific point of view? I think there can be no doubt on the subject. Father Wasmann has pointed out the differences between man and beast, but he has not laid sufficient stress on the great similarities existing between them. The differences are 'in every respect mere trifles in comparison with the marked resemblances.'

> All who heard or have read my third lecture will remember that I laid stress on the resemblances and said: 'It will occur to no scientist to deny that from them certain general arguments in favour of man's descent from beasts may be derived.' Professor Dahl has not proved that the differences which I emphasised are 'in every respect mere trifles in comparison with the marked resemblances.'

Professor Dahl continued thus: 'We are not bound to start from the scientific point of view; the *theological standpoint* will serve us equally well.

If we assume the existence of an almighty Creator, it would seem very strange for this Creator to have made man so completely after the pattern of beasts, if He had created him independently of the animal world. The audience will remember the skeletons, shown them by Father Wasmann, of an ape and a man respectively. The bones are with a few exceptions all identical, differing only in shape. The constituents of the bones and of the other parts of the body show that man and ape bear a remarkable likeness to one another even in the smallest details. This likeness would be *inexplicable*, if we were to assume that an almighty Creator created man as the highest product of creation, *quite independently of the animal kingdom.*'

This argument of Dahl's is *the chief evidence* brought forward from the philosophical standpoint in favour of the descent of man from beasts, and it behoves us to examine it closely and ascertain its real force. Its vulnerable spot is to be found in the words '*independently of the animal kingdom.*' It is undoubtedly true that the great similarity between man and the higher animals suggests that they cannot have been created 'independently of one another,' if thereby we mean to imply that their resemblance is *purely accidental.* This resemblance must certainly be due to *the laws governing the evolution of both,* and so far Dahl is undoubtedly right.

We cannot deny that theoretically there is a connection between the creation of man and of beast, *but we have to determine how far the connection is real.* From the resemblance between man and the higher mammals only one fact can be directly deduced, viz. *that the individual laws governing the evolution of both are based on the same design.* The advocates of the theory of evolution go a step further and say, that *if* the human race has *a history,* we must assume that in its various stages *this history resembles that of the higher animals.* But it *by no means* follows that the histories must be *one and the same,* and that man is descended from beasts. Nägeli, whose remarks on this subject are quoted with approval by Oskar Hertwig (*Handbook of Comparative and Experimental Evolution of Vertebrates,* iii. p. 171, 1906), says, that it is quite conceivable for ape and man to 'stand in no genetic connection with one another and to have distinct lines of ancestry, but this does not hinder the ancestors of both from resembling one another more closely than their modern descendants do, as we can think of the lines only as divergent. . . . We cannot question the fact that primitive cells, spontaneously generated under the same conditions, but independently one of the other, must give rise to similar organisms, provided that their descendants are subject during evolution to similar conditions for a similar length of time.' What Nägeli says here with reference to his theory of spontaneous generation is no less applicable to the theory of creation. If the human race has a history, it need not be *identical* with that of the higher animals, but it may have been due to *primitive cells, resembling one another indeed, but nevertheless essentially different. On these lines all the similarities between man and beast might be satisfactorily explained, without our being forced to have recourse to the theory of man's descent from beasts.*

Professor Dahl continued:—'Father Wasmann laid great stress upon *man's intellectual faculties.* He said that in beasts we could observe only the lower powers, whereas man possesses the higher faculties in addition. I should like to ask him how matters stand with young children. They have only the lower psychical powers, and we can trace in them the gradual development of the higher faculties out of the lower. I do not see why we may not assume that the higher psychical powers have been evolved in just the same way in the case of animals.'

I answered Professor Dahl's question in my closing speech. We see every day that the capacity of thought develops gradually in a child.

This is explicable on the ground of the unity of the human soul, whose higher intellectual faculties develop by aid of the lower sensual powers. No instance, however, has occurred of a young ape's ever becoming a reasoning old ape, for in this case the psychical principle is incapable of those higher mental functions which alone are properly described as intellectual activity.

Professor Dahl next expressed a wish to say a few words on *the origin of the first organisms.* 'It is true that we are still unable to *form any idea* of the origin of these first organisms, and this is probably due to our proceeding from a wrong starting-point. We see *highly developed* organisms, and we assume that the Protozoa, which seem to us low in the scale of development, must have been the primary forms. But our present protozoa, with *their extremely complex protoplasm,* are highly developed organisms, like man and the mammalia. If we accept the theory of spontaneous generation —and Professor Dahl thinks we are forced to do so— 'we are driven to it by scientific considerations'[1]— we must imagine these first organisms as *utterly simple.* This is not a question of possibility, but it is very probable, for, as has been already pointed out (by Plate), these same elements which compose

[1] Dahl must have intended to say, 'by considerations of natural philosophy,' for the results of biological research all tend to disprove the hypothesis of spontaneous generation.

the organisms occur in the inorganic constituents of the earth.'

> Two things may be said against this assertion as to the extreme probability of the spontaneous generation of very simple primary organisms: (1) From the fact that the same elements, which compose the organisms, occur also in inorganic nature, we may infer that the first simple organisms were formed *out of inorganic matter*, but not that they formed themselves from it, as the theory of spontaneous generation maintains.[1]
>
> (2) It is of no advantage to the theory of spontaneous generation to assume that the first organisms must have been very simple. The point is that *they must have had life*. Whether the spark of life showed itself first in some hypothetical 'autoblast,' or in a real 'primary cell,' is quite immaterial; it must have shown itself somewhere.[2]

The speaker went on to say that it would be possible for him to touch upon many differences between his opinions and those of Father Wasmann. He would, however, refer only to *the theological aspect* of this question. 'We hold,' he said, 'two different opinions. Father Wasmann thinks *that*

[1] Cf. my remarks on Plate's speech, p. 99, etc.

[2] See my remarks in the 3rd edition of *Biology and Evolution*, p. 202.

God intervened three times in the course of creation; I maintain that only *one action* was necessary. In his first lecture, Father Wasmann showed that we should have but a poor idea of God, if we thought it needful for Him to interfere at every turn, in order that all things should follow their normal course. I *fully agree* with Father Wasmann on this subject, and I think that my conception of God, as intervening *only once*, is a far higher conception than that of Father Wasmann, who imagines God to have intervened *three times*.'

> If Professor Dahl is in a position *to prove* that we need not assume any subsequent intervention on the part of the Creator, in order to account for the origin, first of life, and then of the intelligent soul of man, I will gladly accept his view of God, but he has not supplied us with any such proof.

With reference to *the question of design* in nature Professor Dahl said: 'I wish to state in general terms that *the theory of selection* is the only one which can take the place of that of beneficial purpose. On that one point I am in complete accord with Father Wasmann.' [1]

[1] There is a misunderstanding here. Neither in my second lecture nor in my published works had I any intention of saying that the theory of selection *ought to take the place* of that of beneficial purpose, but only that the former ought *to be the complement* of the latter. Unless we presuppose the existence of some immanent directive principle, *selection has no object* at all, and I have often pointed this out. Cf. my remarks upon Plate's speech, p. 106, etc.

'Father Wasmann has also referred to the advance made by Darwinism or the theory of selection, but he says that he believes this advance will eventually lead to the establishment of quite a different theory. It is true that the theory has advanced, but unfortunately Father Wasmann has failed to perceive in what direction, otherwise he would have known that *his amical selection*, the theory which he considers he has established in *opposition to that of selection*, and which led him to deny the universal applicability of the theory of selection, was thoroughly explained twenty years ago, according to the principles of the theory of selection. It is certain that no logical contradiction exists, and that is all with which we are now concerned.'

I subsequently asked Professor Dahl to what work he had referred in these words. He kindly informed me that he had been referring to his 'Versuch einer Darstellung der psychischen Vorgänge in den Spinnen' (An Attempt to explain the Psychical Processes in Spiders), which is the second article in the *Quarterly Journal of Scientific Philosophy*. Avenarius ix. (1885), pp. 162-190. In the third section of this article Dahl speaks of æsthetic sensations (p. 184 *et seq.*), and discusses the sexual selection among spiders, which has given rise to definite characteristics possessed by the males (excessive development of the eyes) which are not of any use, but actually detrimental in the struggle for existence. I find no allusion here to *amical selection*, which differs from sexual selection, and was advanced by me, not earlier than 1897, as a theory capable of affording an explanation of the development of the true relation in which inquilines stand to their hosts.[1] I cannot therefore understand how my 'theory of amical selection was fully explained twenty years ago according to the principles of the theory of selection.' Moreover, to discover whether a *logical* contradiction exists is *not* our sole object on this occasion, but we have also to ascertain whether there is an actual discrepancy between the theories of *amical* and *natural* selection.

In the work quoted above, published in 1885, Dahl proved the existence of a similar real discrepancy between *sexual* and *natural selection.* To this extent therefore there is a certain likeness between Dahl's earlier and my later work. But it was not the theory of amical selection alone which is supposed to have led me to regard Darwin's theory of selection as merely a *subordinate* factor. (Cf. my second lecture, pp. 41, etc.)

Professor Dahl concluded his speech with these

[1] 'Zur Entwicklung der Instinkte' (Development of Instincts) in the *Proceedings of the Zoological and Botanical Society*, No. 3, pp. 168-183. Vienna, 1897.

words: 'I think that we shall do well to discuss in writing the points on which we differ.[1]

'I should like to insist upon only one more point —viz. that our arguments, or rather the differences that exist between us, are not personal. We may be friends, although we may stand in strenuous opposition to one another. *One bond* unites us all, *one quest* inspires us all, there is *one aim common* to us all,—*we are all in search of truth*!'

IV. Dr. Friedenthal's Speech.[2]

'The scientist Wasmann in his last lecture, on the application to man of the theory of Descent, criticised my scientific work in a manner that calls for a short reply on my part.

'In a passage which he quoted,[3] and from which

[1] I should be glad to fall in with this proposal, especially as the closing portion of Dahl's speech has convinced me that we may reasonably hope for increased mutual understanding. The conciliatory tone which distinguishes this speech made itself felt also in his work, published in 1886, on *The Necessity of Religion: an Ultimate Result of Darwin's Teaching*. Although I cannot agree with most of the opinions expressed in this work, I feel bound to recognise the author's good intention.

[2] If the reader will kindly compare the remarks in my third lecture on the subject of Friedenthal's research work on blood, he will be better able to appreciate the agreement between this speech and my statements.

[3] The quotation was made on Rössle's authority; cf. the *Biologisches Zentralblatt* for 1905, No. 12, p. 422: 'At the Anthropological Congress at Greifswald in 1904, Uhlenhuth reported having observed a positive reaction on the part of human antiserum with the blood of pithecoid apes. Friedenthal, too, mentioned having quite recently obtained a positive result with the blood of lemuroids.'

he inferred that I had lately lost confidence in the results of my investigations regarding blood, I merely pointed out what precautions must be taken to avoid certain sources of error. There was no question of *a retraction* of the results of my work, but only of a reference to the avoidance of error.'

However, it was not this criticism of his works that caused Dr. Friedenthal to address the meeting, but he wished to point out that all present were under the impression, that as soon as Father Wasmann began to speak of the origin of man, he spoke as a *dilettante scientist*, in sharp contrast to the Father Wasmann, whom they all knew as a specialist devoted to research work on the subject of ants.

By calling me a 'dilettante scientist' in my dealing with the origin of man, Friedenthal seems to mean that I am not a specialist. I was quite unaware that one, who is not a specialist, is supposed to be incapable of criticising *the conclusions* at which a specialist in that field of research has arrived, provided that he takes pains to obtain information regarding the subject and the results obtained by others.

Dr. Friedenthal did not prove in his speech that I had criticised his investigations in a *dilettante manner*; on the contrary, he allowed that I was right in saying that a chemical and physiological likeness between two kinds of

> blood must not, without further consideration, be regarded as proving *blood-relationship* in *the sense of having a common origin.* Apart from his experiments on the reaction of blood, Dr. Friedenthal himself is as much a 'dilettante scientist,' *i.e.* a non-specialist on the subject of the origin of man, as I am.

Friedenthal expressed a wish to distinguish clearly the realms of *the natural* and *intellectual sciences* respectively. He declared the former to be concerned with processes of motion, and the latter with conceptions and ideals. In dealing with the natural sciences, he said, the scientist was free and unfettered, and could establish his results regardless of historical evolution and of dogmas. But in dealing with conceptions or ideals, there would always be warfare, for here there was an absence of the proofs which are a preliminary condition for the attainment of definite results. Consequently, in the realms of ideals, the attainment of results did not depend upon so-called *proofs*, which no one could fail to accept, but rather upon personal feelings. As to his own attempts to *prove community of origin from blood-relationship*, this was a matter in which there was no *logical evidence* acting irresistibly upon each individual. The speaker considered that we could not bring forward such conclusive evidence of origin, even in the case of a child of some particular married couple. '*The question of the origin of man stands upon exactly the same level.'* Here too it is not a matter of proofs, which are to convince any chance person, but of indications, which may be of assistance to any one who studies these questions, for the *subjective evidence* would be so irresistible that the questions would vanish.

In reply to these statements, I should like to suggest the following considerations.

First of all, in order to avoid sources of error in dealing with the natural sciences, it is of great importance to keep in view, not only the evidence which the specialist has himself collected, but also the historic development of his particular department of research, and the assured results attained in other departments. Secondly, the province of conceptions is not co-extensive with that of ideals. The former, like that of the natural sciences, is subject to strict logical laws of thought, whereas in the latter, the ultimate decision, as to which ideals are worthy of man, rests with the reason. After Friedenthal had just declared that, in the case of the natural sciences, irrefutable evidence was always a necessary preliminary to the attainment of results, it was certainly interesting to be told that logical proofs altogether broke down *in dealing with questions of descent.*

Even the question whether a man were really the son of his parents or not, would, according to Friedenthal, belong to the province of concepts, or even to that of ideals. I can scarcely believe that his fellow-citizens would all agree with him in this respect. As far as I am concerned, 'subjective evidence,' which is only a matter of sentiment, would not at all suffice to answer all questions arising about the descent of man. I feel the need of '*objective evidence*' possessing at least a high degree of probability.

As the speaker proceeded, he said that he felt bound to point out how unfairly the scientist Wasmann had treated *the morphological law of motion.* Friedenthal himself had only followed many other scientific men in claiming no particular place for man among the mammals of the zoological system; he was content to class man and other apes together in one subdivision of mammalia, but in the newspapers this opinion had been branded as atheism by the Church. He said that he could not imagine any view of God so utterly opposed to scientific research, as that a scientist, who insisted upon the resemblance between man and other apes, should therefore be accused of *atheism.* Yet such was Father Wasmann's point of view.

In answer to Dr. Friedenthal I should like to say, that in my third lecture I was careful to

> show how man, with respect to his body, is the most highly developed type of mammal; but I regard it as morphologically wrong *to place him with apes in the same order of our zoological system.* Therefore it seems to me unsuitable to speak of a 'resemblance between man and *other* apes.' This was my standpoint also in the 3rd edition of my work on *Biology and the Theory of Evolution* (chap. ii.). But neither there nor in my lectures did I ever accuse Dr. Friedenthal of *atheism,* because he classed man, with respect to his body, with apes, so as to form one order in our system. The speaker is therefore not justified in calling this 'Father Wasmann's point of view.' In my opinion the *mental divergency* constitutes the chief difference between man and beast; the bodily differences are of less importance.

Dr. Friedenthal arrived at last at his chief argument. '*What is proved by the researches made into blood-relationship? Nothing, but what Father Wasmann himself said, viz. a chemical resemblance on the part of two individuals, whose resemblance could not be established without these investigations.*' This was, however, in the speaker's opinion, a new point, saving people the trouble of raising further questions and of doubting whether they were dealing with different individuals (!) It had been proved that morphological resemblance was, as a rule, enough to establish relationship, and in the course of sixteen thousand experiments, carried out by Nuttal and his pupils, no distinction had been found between morphological and chemical resemblance.[1]

'Thus if Wasmann means that nothing beyond this has been attained by these experiments, I agree with him ; but a new aspect of relationship has been revealed, and that is what my experiments were intended to demonstrate.'

> I accept this correction of Dr. Friedenthal's with great satisfaction. (Cf. my closing speech.) A resemblance between different kinds of blood must not be confused with a *blood-relationship* in the sense of *a community of origin*, as has been very widely done. Friedenthal's former dictum : 'We are not merely descended from apes, but we are ourselves genuine apes,' was not intended to bear the meaning which we felt its literal interpretation required. We have this on his own authority, and I am perfectly satisfied.

Friedenthal next went on to discuss *the peculiarities of the human soul*, and said that on this topic he agreed more closely with Father Wasmann than the previous speakers had done. The soul existed in the world of concepts or ideals, which

[1] In my third lecture I quoted Rössle, who proves that in many cases morphological and chemical resemblance do not coincide.

could not be brought into connection with the sense-organs or their perceptions. '*I have no reason for suspecting the existence of such concepts or ideals in any other living creature, not even in the highest animals.*' But, he went on to say, he had no more reason for suspecting the existence of concepts or ideals in a new-born child, or in one still unborn, and only in process of formation.

> The recognition made by Dr. Friedenthal of my psychological opinion (as opposed to Bölsche, von Hansemann, Juliusburger, and other speakers) was a pleasant surprise. That in man the higher intellectual faculties can enter into a state of activity only after the lower faculties with their nerve centres have been developed, is explained in my remarks upon Dr. Juliusburger's speech.

'We shall be asked,' said Friedenthal, 'how we know anything of *the soul of an animal*; whence we learn what animals feel and what ideas they have. Our knowledge is based upon conclusions derived from analogies. We cannot actually know what an animal thinks, but its movements and general behaviour force us, bearing in mind its previous history, to conclude that *an animal has no concepts and ideals.* And Psychology teaches us why man is so radically unlike beasts—because man alone possesses *speech,* from which concepts and ideals can be formed.'

This fact, according to Friedenthal, serves as a proof, '*that in man is something peculiar, which marks him off from all other living creatures, and I agree with the scientist Wasmann on this point which distinguishes man, as we know him, from all other living beings.*'

> Speech is undoubtedly an important *aid* to the formation of concepts and ideals, but it is *not their primary cause.* This must be sought in the human intelligence. (On this subject cf. the 3rd edition of my work on *Instinct and Intelligence in the Animal Kingdom.* Freiburg i. B., 1905, p. 92 *et seq.*). Otherwise I can agree with these remarks.

Dr. Friedenthal considered that Father Wasmann had been illogical. At the outset he had declared that scientific research could never clash with religious conviction. Dr. Friedenthal, too, could not imagine a religion capable of being upset by any scientific fact whatever. Father Wasmann's want of logic showed itself later on, when, instead of speaking as a scientist, he became a partisan, taking up his position in consequence of religious considerations.

> Dr. Friedenthal would be at a loss if he tried to prove that in my lectures any *religious considerations* led me to decide against the hypothesis of spontaneous generation, or against the intellectual evolution of man from

beasts, etc. Purely scientific and philosophical considerations determined my decisions.

Finally, Dr. Friedenthal recurred to his conviction that, in classing men with apes, he had done nothing which could offend the susceptibilities of a layman or of a religious person. He pointed out that the pious Linnæus, who has never been accused of being anything but a good Christian, put man into the same class as apes.

The 'pious Linnæus' seems not to have classed man with apes, for he calls him *Homo sapiens*, not *Simia sapiens.*

Once more I wish to say clearly that Dr. Friedenthal by no means deserves to be called an atheist because he thinks that man, *with respect to his body, belongs in the same systematic order as apes.* Moreover, he expressed himself more correctly than any of the other speakers, on the subject of the mental difference between man and beasts.

Dr. Friedenthal concluded his speech by uttering another protest against confusing scientific problems with religious questions. I thoroughly agree with him on this point, and, above all, I think that a wrong use of scientific results is made, when they are employed as weapons against Christianity, after the fashion of Monism under Haeckel's guidance.

V. Professor von Hansemann's Speech.

The speaker began by asserting that Father Wasmann had said in one of his lectures, he believed, in the first of all, that natural science by itself was not entitled to express an opinion regarding *the problem of the evolution of animals.* Professor von Hansemann, however, believed that natural science alone had a right to express any opinion at all on this subject; it had therefore been *a great mistake* to drag religion, theology, and Christianity into this whole discussion. These things ought to be absolutely excluded from scientific deliberations, and, if this were done, it would be a much easier and speedier task to arrive at an agreement, for in his opinion *it did not affect the questions* under discussion at all whether a man had any religious sentiment, whether he was interested in theology, or whether he upheld the Christian theory of the position of man in the universe.

Before discussing this introduction to von Hansemann's speech, I must make it clear that the first sentence contains a *manifest error,* which might easily have been avoided. He imputes to me a statement to the effect that the evolution of animals was not a problem within the scope of natural science, *whereas I carefully proved the exact contrary* in the first of my three lectures. The speaker confused the evolution of *animals* with that of *man*, with regard to which I showed in the third lecture that it was not a *purely* zoological problem. Von Hansemann is, however, perfectly right in saying that it was a mistake to drag religion and theology into the discussion of my lectures. The result of so doing has not been the refutation of my scientific and philosophical opinions by my opponents, but the transference of the whole discussion to a region lying beyond the province of a scientific conference. Unhappily von Hansemann was not able to avoid the mistake which he pointed out, as his subsequent remarks clearly show.

'Religion,' said the speaker, 'is a matter of faith, whereas natural science is the quest of knowledge. What would become of us, if we had to behave like Father Wasmann, and say: "As long as the supreme authority has not decided anything, I cannot venture to express my opinion." If Copernicus had waited until the supreme authority had stated its views, the sun would still be moving round the earth. It is impossible in our day to carry on scientific research thus.'

This remark is also the outcome of a misapprehension on the part of the speaker. Even if the ultimate decision of what is admissible from the theological point of view rests with the highest ecclesiastical authorities (cf. Lecture III. p. 54), we are by no means debarred from previously expressing an opinion on the evolution theory, as this is a question of a very complex character. Consequently I stated my views upon it in my lectures, in spite of Professor von Hansemann's scruples. His remarks on Copernicus proceed also from ignorance of the true facts. It is well known that Copernicus in 1543 dedicated his work *De Revolutionibus Orbium Caelestium* to Pope Paul III., who accepted the dedication. The work treated of an astronomical problem, not of a theological one, so von Hansemann's reference to Copernicus proves nothing at all.

The speaker passed on to the subject of Virchow, and expressed his satisfaction that Father Wasmann had recognised him as an authority in criticising the *Pithecanthropus erectus* and the Neandertal man. He said that in 1874, Wasmann's colleagues had spoken very differently about Virchow, when he would not commit himself on the subject of the impression of the stigmata upon Louise Lateau; at that time he had been slandered and overwhelmed with abuse.

It is difficult to see what the stigmata of Louise Lateau have to do with my Berlin lectures. They certainly need not have been mentioned at this discussion.

Von Hansemann proceeded to criticise the Christian view of the position of man in the universe. He asked what this view really was, and went on to say: 'We must remember that, long before there was a Christian view at all, there were scientific theories, which are constantly being proved to have corresponded very well with facts as known to men. The Greeks, the ancient Egyptians, the Hindoos,[1] and the Chinese had excellent theories of a scientific character. Therefore all this has nothing to do with religion or with Christianity. If these questions depended upon the Christian theory of life, what would become of the Japanese, who enjoy at the present day the highest scientific education? Because they are not Christians, are they to be excluded from considering the theory of evolution and other similar problems? We ought to limit ourselves here strictly to the scientific standpoint, and proceed from it in dealing with these questions.'

Von Hansemannn is here plainly confusing the *theory of nature* with *the theory of life*. The theory of the atomists, for instance, who explain

[1] It is a pity that the Professor did not more closely compare the fantastic myth of the Creation, as told by the Hindoos, with the simple and dignified account of it which is contained in the Bible.

all the processes of Nature from her material aspect, as being a system of atomic motion, is, in itself, no theory of life. It becomes such through bold, and in this case false, *philosophical generalisations*, when the assertion is made that nothing can exist except systems of moving atoms. That atomistic philosophy, as *a theory of nature*, has nothing to do with either the heathen or the Christian theory of life, is obvious. Von Hansemann ought to have distinguished his ideas more clearly, before undertaking to enlighten his audience on the relation existing between the Christian theory of life, and the scientific doctrine of evolution among the Japanese.

The speaker continued :—' Father Wasmann thinks that the *numerous problems*, which force themselves upon us, can be reduced to one great problem.[1]

' But even he does not succeed in thus reducing them, for he soon discovers a new problem, viz. that of *vitalism*. The vitalist is a man who has devoted himself more or less to the consideration of scientific questions, and if he has not succeeded in solving them, he has said to himself : " There is still something that cannot be explained, because

[1] Von Hansemann seems to be referring to the problem of *creation*. But have the origin of the first organisms and vitalism nothing to do with this problem, as he alleges ?

I cannot explain it," and thereupon he straightway becomes what is called a vitalist.'

The opinion expressed by von Hansemann on vitalism and the vitalists is injurious to no one but the critic himself. Scientists such as Driesch, von Bunge, Reinke, etc., have certainly advanced the interests of science by their vitalism, far more than von Hansemann has done by his antivitalism. Because he does not understand vitalism, it does not follow that vitalism is nonsense.

'The vitalist maintains that there is a principle of living force underlying all the vital processes. He is unable to say what this "living-force" principle is, and so he has recourse to the transcendental, thus rendering it impossible for him to carry on any investigations in natural science, for such investigations, as Helmholtz showed, postulated that the questions propounded should be theoretically capable of solution. As soon as any one adopts the vitalistic standpoint, the question ceases to be theoretically capable of solution.'

Von Hansemann here simply assumes as *self-evident*, that a *theoretical* solution of problems in natural science must necessarily be a *purely mechanical* solution. But as, according to the vitalists, the phenomena of life are *not* explicable in any purely mechanical way, they are therefore 'theoretically incapable of solution' as far as vitalism is concerned. This argument against vitalism is not a success logically, for it assumes, as already proved, what it had undertaken to prove. The speaker next referred to teleology or purposive action (lit. efforts to attain an end—'Zielstrebigkeit'). 'We may,' he said, 'define expediency by the aid of philosophical subtleties or in any other way, but we cannot avoid the conclusion that expediency and the doctrine of adaptation to purpose ("Zweckmässigkeit") are as much alike as two pins.'

The underlying reason of this similarity seems to have escaped the critic. It is, of course, that adaptation to purpose is the motive of expediency.

'Expediency,' continued von Hansemann, 'comes to this: that we conceive all things to be *as well as possible adapted* to their surroundings.'

Von Hansemann here betrays the same mistaken idea of absolute expediency on the part of all living creatures, as we corrected in the remarks on Plate's speech (cf. p. 106). Consequently von Hansemann has not succeeded any more than Plate, in proving anything against real expediency, which corresponds only to the normal and average circumstances of organisms.

We now come to the evidence tending to show that organisms are not well adapted.[1]

Pathologists have, according to von Hansemann, the best right to be heard here, for we constantly light upon instances of defective adaptation, especially in men ; for instance the teeth, which cause pain when they are cut, when they are changed, and when they decay. Elephants and rodents are much better equipped in this respect, for their teeth are constantly growing ; and tortoises are better off still, for they possess a horny substance which takes the place of teeth, whilst granivorous birds have a gizzard. The human teeth are therefore a '*clumsy* arrangement.' Would not the Professor nevertheless regret it, if he had no teeth ? They certainly answer their purpose, though perhaps not in such a way that no better method could be imagined.

Further, the methods of reproduction among mammals are, according to von Hansemann, '*altogether inexpedient*'—an arbitrary assertion, which he has not proved to be true. He derives his chief argument for the absence of expediency from pathology, which he regards as consisting of a chain of instances of inexpedient devices, of faulty adaptation of the individual to his surroundings.

[1] Haeckel and other materialists have drawn up a very long list of dysteleologies, or instances of non-beneficial action. For a criticism of it see the *Apologetische Vorträge* (Apologetic Lectures), published by the Volksverein for Catholic Germany, No. 2, 1907, p. 125 *et seq.*

> Von Hansemann has completely overlooked the fact that pathology is the science which deals with man *as diseased*, not with man as healthy, and even at the present day *the healthy* must be regarded as the normal condition for man.

The speaker next brought forward an argument against his own statements. 'In apparent absence of purpose or inexpediency, there is concealed real purpose or expediency, inasmuch as it renders possible a *selection* from among the various individuals. Selection, however, is a "makeshift on the part of nature," and affords no evidence of design in nature. Nature somehow attains her end, but she is not acting with design.'

> We are impelled to ask the speaker, how it is then possible for nature *with no design* to exist at all. If there were no forms conforming with that design, what would selection find to select? This shows plainly what I pointed out before with reference to Plate, that design is the necessary condition which renders selection possible. Selection is only a *subordinate factor*, which *presupposes* and *supplements* the design immanent in organisms.

Von Hansemann then passes on to *the rudimentary organs* in man. He thinks they may be regarded as 'dwindled organs,' the remains from previous phylogenetic periods. He refers to Wiedersheim's work on the subject, for Wiedersheim has proved this fact in a most admirable manner—so admirable a manner, we may add, that as long ago as 1892 Hamann wrote a most crushing criticism upon his fantastic opinions.[1]

Von Hansemann lays particular stress upon the vermiform appendix as an instance of a rudimentary formation in man. It seems to him improbable that inflammation of the appendix occurs less frequently among savages than among ourselves, but he wishes for information on the subject.

> Darwin and his followers believe this vermiform appendix to be of great importance in determining the history of the human race, and so I may make a few remarks upon it here. (Cf. my third lecture, p. 65.)
>
> Prominent pathologists of the present day believe that the appendix, with its abundant lymphatic tissue, is of the same kind of use to the intestine, as the tonsils are to the palate. Formerly we used to hear that the appendix was of no use at all, and it was condemned simply as being 'rudimentary,' but now it is generally recognised that it probably has some definite function to perform with reference to the intestine, although the nature of this

[1] See my *Biology and Theory of Evolution*, p. 451 *et seq.*

function has not yet been ascertained. Some years ago, Ribbert, a great pathologist, writing in Virchow's *Archiv*, expressed the opinion that an obliteration of the cavity of the appendix, occurring during life, was *a typical instance of the process of involution.* If this were true, the appendix would seem to be really a rudimentary organ. But in 1902 it was shown by Dr. Joseph Koch that *these obliterations are to be regarded solely as consequences of previous inflammation.* (Observations on chronic or recurrent perityphlitis based upon two hundred operations, *Archiv für klin. Chirurgie* (Records of Clinical Surgery), vol. lxvii., 1902, Part II.) An exact proof of the accuracy of this theory was given in 1904 by Professor L. Aschoff, in the *Proceedings of the German Pathological Society* (p. 246, etc.), in an article on the topography of appendicitis. Professor Rotter, of the Hedwigskrankenhaus in Berlin, placed at his disposal for purposes of investigation one hundred and three vermiform appendices. The work already mentioned by Dr. Koch, who was then assistant physician at the same hospital, is based upon the same materials. According to these specialists, the vermiform appendix of the cæcum closes up only in consequence of morbid appearances. Under normal conditions in a healthy person, we may infer, judging from its gland tissue, that it has to perform the physiologically important function of an accessory or collateral gland of the intestine. What exactly this function is, remains still to be discovered. At any rate, this theory has weakened the old opinion that the vermiform appendix was a rudimentary organ. With reference to the stress laid by von Hansemann upon pathological phenomena in man, in order thereby to establish his dysteleology (or theory of absence of design), I may here point out, that as long ago as 1897, one of our most eminent pathologists, Professor G. Bier, von Bergmann's successor in Berlin, writing in Virchow's *Archiv*, propounded and established the thesis that inflammations are not *instances of inexpediency*, but are, on the contrary, *beneficial prophylactic devices* on the part of an organism to rid itself of bacteria or other injurious matter that may have penetrated into the system.[1]

An interesting discussion upon Dr. Bier's theory took place at the 35th Congress of the German Surgical Society at Berlin on April 7th, 1906. (See the *Proceedings*, pp. 220-265.) Against all objections Professor Bier maintained his thesis with success, showing that *inflammation is a process beneficial to the organ-*

[1] On this subject cf. Professor G. Bier's work, *Hyperæmia as Means of Cure*, Leipzig, 1907; also the article entitled 'Bier's Treatment of Hyperæmia' in the supplement to the *Allgemeine Zeitung*, 1907, No. 89, pp. 107-109.

ism. These remarks will perhaps suffice to prove that von Hansemann's views on inexpediency in human pathology are, to say the least, very one-sided.

Let us now return to the speaker who alluded to my opinions on *comparative Psychology.* He said: 'On this subject, though it is less evident in his lectures than in his writings, Wasmann quits the firm ground of assured scientific facts, and passes over to *definitions,* but he formulates these definitions in such a way, that no other conclusion is possible, than that the things in question are peculiar to man, and are not possessed by beasts. This is how he proceeds in dealing with reason as opposed to instinct. He might just as well prove that man alone has a brain, and beasts have none; for, if he defines the brain as the nervous organic centre, not only man, but many beasts have brains; but if he says the brain is an organic centre, contained in a skull, having definite functions, and weighing so much,—then man alone has a brain and beasts have none. *This is how he manipulates his evidence to show that only man has reason, and beasts merely instinct.*'

I have printed the last sentence in italics in order to draw more attention to it. Professor von Hansemann seems not to have read my writings to which he refers;—he is probably alluding chiefly to my work on *Instinct and Intelligence in the Animal World* (3rd ed., Freiburg i. B., 1905)—as otherwise it would not be possible for him to give so superficial and so erroneous a description of my methods of proof. Without a previous clear definition of the ideas involved, it is quite obvious that it would be futile to debate whether animals possess reason or not. Only by means of philosophical consideration of each point, can we arrive in this subject at any intelligible result. Critics who have thought more deeply, such as Professor Emery, have expressly recognised the fact, that my chief merit in the treatment of this question lies precisely in my clear definitions of the ideas involved.

At the conclusion of his speech von Hansemann referred again to the 'almost absolute agreement' existing between the pathology of man and that of beasts. He said that only differences of degree existed, not of principle.

I failed to understand what inference was to be drawn from this remark that could militate against the essential difference between man and beast, with regard to their *mental* equipment —especially as the speaker did not touch upon diseases of the brain.

VI. Count von Hoensbroech's Speech.

This speaker came forward having under his arm three thick books, viz. Wasmann's *Biology and Theory of Evolution*; *Canon Law*, by Father Wernz, the present General of the Society of Jesus; and the *Index of Prohibited Books*, by Father Hilger, S.J.

Count von Hoensbroech spoke for twenty minutes, but, without referring to the theory of evolution, he talked about the Russian censorship of books exercised by the Jesuits, about the Roman Index of prohibited books, about the Syllabus, about Canon Law and about the Vatican Council, and from these sources he tried to prove *a priori* that Father Wasmann, being a Jesuit and a good Catholic, could not be a 'free scientist.' [1]

In all this there was not the slightest reference to my Berlin lectures from the scientific point of view. The whole speech was nothing but one of those tirades against the Catholic Church, which we

[1] According to this speaker, there has, in fact, never existed among good Catholics 'an explorer in the domain of natural science.' He seems unaware of the fact that the great Copernicus was a Catholic canon of Frauenburg. The unscientific character of von Hoensbroech's remarks on the Roman Index has already been pointed out in the *Kölnische Volkszeitung*, 1907, No. 498, in an article signed 'Also a Representative of Science.'

The sharpest condemnation of von Hoensbroech's behaviour at the evening discussion was pronounced by a Protestant critic, who does not personally accept even the theistic point of view. His article, entitled 'A New Scientist,' and signed Pilatus (Dr. V. Naumann), appeared in the *Deutsches Volksblatt*, 1907, Nos. 72-74.

have heard *ad nauseam*, and it was obviously out of place, for Professor Waldeyer, who was presiding, had said when opening the discussion: 'I have taken the chair assuming that the proceedings are to be strictly limited to calm, genuine, and scientific expressions of opinion.'

This was the reason why, in my closing speech, I simply disregarded Count von Hoensbroech's remarks as being out of order.

VII. Mr. Itelson's Speech.

In spite of the benevolent intention of the speaker to prove, from an historical standpoint, that there was something pleasant about Father Wasmann's appearance as a scientist, notwithstanding 'the oppression of the Church,' this speech too will not be reported in detail, for it, like the one that preceded it, had nothing to do with the subject treated of in my lectures. It consisted of historical platitudes about the presumable decay and crumbling away of the rock of the ecclesiastical theory of life, as the waves of science encroached upon it. Unlike von Hoensbroech, who had simply denied me the possibility of 'free research,' Itelson thought that he might describe me as a fragment of the Christian rock, already in process of disintegration. My answer on this topic was given him in my closing speech.

The remarks of the eighth speaker were really to the point, and therefore they will be reported fully with critical remarks, especially as, in my closing speech, I had no time to deal adequately with the eight headings of his argument.

VIII. Dr. Juliusburger's Speech.[1]

'Ladies and Gentlemen,—

'Father Wasmann states it as his opinion, that *zoology* alone is not competent to decide the question of the origin of man; on the contrary, he maintains that *psychology* has the best right to express its views on the subject, and, as constituting an absolute barrier between man and beast, he represents that man alone possesses a *simple soul*, which is the higher part of his intellect.

'On this topic I wish to make the following remarks: 1. *It is a mistake to identify the soul exclusively with the intellect*, the truth being that the foundation of the psychical processes is to be sought rather in the *will* or in the *feelings*. If this truth be recognised, there follows, by direct intuition, that *all living beings are essentially alike*, that there is an essential identity between plants, beasts, and man, apart from secondary distinctions.'

[1] I reproduce this speech verbatim, according to the shorthand writer's report, as Dr. Juliusburger gave me express permission to do so.

Answer to No. 1. It would undoubtedly be a mistake to regard the soul as belonging only to the intellect, but a mistake is made no less by assigning it to the will or the feelings; for the essence of the soul does not consist of its *activity*, whether intellectual, or voluntary, or sensitive, but the soul is *the efficient cause* and *the permanent subject* of all these phenomena. Nevertheless we are right in deducing, from the unique character of these psychical *processes*, the unique character of their *efficient cause*, and therefore the unique character of the soul.

If it were possible to set the desires and impulses of beasts and plants on a level with acts of human volition, we might be able to maintain that there was *an essential likeness* between man, beast, and plant. But as long as this has not been done (and Dr. Juliusburger does *not prove* its possibility), even from the point of view of voluntarism, this conclusion is not admissible. We must indeed assume, that there is an *interior principle* of perception and desire, but we must assume its existence *only to the extent* in which it manifests itself *exteriorly*. In plants we find *no evidence* at all of perception and desire, and in beasts only such as shows them to be *essentially different* from human acts of volition and thought, though not from the lower acts of sense.

Thus it is to man alone that we may ascribe

an interior principle underlying acts of volition and thought, *i.e.* a *simple, intelligent soul.*

2. '*There is in man no simple soul.* Analysis reveals to us sensations or perceptions of a simple kind, impressions of such perceptions on the memory, which form the basis of ideas, and connection of these ideas, so as to make complex pictures in the imagination. But the concrete and the abstract ideas have one common source, viz. sensation or perception. The material substratum of the connection of sensations and ideas with one another is to be sought in what anatomists call the associative fibres. Perceptions and ideas are closely connected with feelings and impulses. This associative theory, thus briefly indicated, *does not admit* of our assuming the existence of a simple soul.'

Answer to No. 2. The necessity of the presence of *a simple soul* in the human body cannot be proved *directly*, either by introspection or by objective experiments, but only by *deduction.* We must deduce the existence of a soul as the psychical principle from each single psychical act. The *simplicity* of the soul is deduced from the existence of a *simple personal consciousness* or self-consciousness, as also from the psychological analysis of the powers possessed by man of forming concepts, judgments, and inferences.

Sensation, sense-perception, and imagination *do not by any means account for* the whole of our psychical life on its perceptive side. General concepts, judgments, and inferences are far higher components of our knowledge, and psychical association does not afford a satisfactory explanation of them, far less does association in the physiological sense, which depends anatomically upon the so-called associative fibres. Our concepts, judgments, and inferences are, of course, built up upon a foundation of sense-perception, but in a way differing altogether from that in which the sense-images are formed, for sense-perception is a real element in these. The old saying, *Nihil est in intellectu, quod non antea fuerit in sensu,* is true only in as far as sense-perception is a preliminary condition to true intellectual activity, and constitutes *the material* from which it is formed. The associative theory, properly understood, and kept within its true limits, is therefore absolutely compatible with our assuming the existence of a simple soul; in fact, it even leads inevitably to this assumption.

3. 'Experiments upon animals and observations of human beings have successfully shown that a great part of our psychical activity must *be referred to a particular region of the cerebral cortex.* I may remind you that our visual perceptions are con-

nected with the occiput, and the control of our actions depends upon definite parts of the brain. Therefore what is regarded as belonging to the *higher intellectual life is not a simple, but a very complex quantity,* and it has been proved that various important parts of it are located in various portions of the cerebral surface. The consciousness of volition may be regarded as the expression of the activity of all the cerebral cortex collectively. But, on the other hand, certain sensations, which we term organic, and which are connected with the activity of the body and of the organs of sense, must be referred to definite regions of the brain. *Everywhere we find a complex, compound soul, nowhere a simple entity.*'

Answer to No. 3. It is certainly clear, and is expressly admitted by us, that a large part of our psychical activity is connected with the cerebral cortex. This is true of all those psychical processes in which the nerves play an essential part, viz. sensation, sense-perception, and sense impulses and desires. (The indirect, external dependence upon nervous processes is granted also in the case of the higher activity of the soul.) To this cerebral cortex belongs everything which our language has derived from the domain of sense—*e.g.* when a word is to be spoken or written, we have images of its written appearance, its sound, and of the movements required for its production. On this is based the localisation of the cerebral functions, as far as hitherto it has been proved, whilst definite psychological processes are connected by their physiological element with a definite brain centre.

This localisation rests upon a physiological division of labour, and proves nothing at all against *the simplicity* of the soul. As the above-mentioned psychical processes, in their actual connection with one another and with sense attention, practically occupy the whole area of the cerebral cortex, the soul must necessarily be present *at every point of the cerebral cortex*, just as it must be present in every part of the body to which it imparts life. It does not, however, follow that *the soul itself is anything compound*, as Juliusburger maintains, for a simple entity can at the same time be present at various points of an extended body.

Hitherto no trustworthy evidence has been adduced to show that the *higher* psychical functions, the *intellectual activity*, strictly so called, can be localised in any definite regions of the cerebral cortex. I agree with K. von Monakow, who says on this subject that we have not yet 'advanced beyond an uncertain groping about on the surface of the brain.' (*Ergebnisse der Physiologie*, iii., 1904, Part II. p. 122.)

The points on the cortex which Flechsig calls 'organs of thought,' have by no means been proved to be such. In his work on Functional and Organic Nervous Diseases, entitled *Grenzfragen des Nerven- und Seelenlebens* (Investigations regarding the limits of nervous and intellectual life), ii., 1900, p. 77, Obersteiner states very definitely: 'We see, in fact, that we can ascribe with certainty to the *known* cortical centres only processes of a more material character.' On p. 78 he describes Flechsig's discovery of the 'organs of thought' as an unsuccessful attempt, assailable from the anatomical, as well as from the physiological and clinical points of view. Thus Dr. Juliusburger asserted more than he was in a position to prove, when he said that, by means of localising the cerebral functions, evidence had been afforded that our *higher intellectual life was 'not a simple but a very compound quantity.'* Only its lower subsidiary processes have hitherto been to some extent regarded as localised. Thus he has proved nothing against the existence of a simple soul, as both the lower and the higher activities of the soul *unite in one simple psychical joint action.*

This answer may be summed up shortly in the following sentence:—*The intellectual life of man, regarded as an accumulation of isolated acts, is naturally not a simple, but a very compound total. But the inner, essential, underlying principle of these acts can be nothing but a simple, intellectual entity.*

4. 'In certain mental diseases, for instance in what is called Softening of the Brain, the most refined mental processes are first affected and destroyed, whilst the so-called lower intelligence is affected later, and is not completely destroyed. Another instance is given in the case of melancholia, where only the most valuable qualities of the soul are affected.'

Answer to No. 4. Juliusburger's reference to the symptoms of softening of the brain is not to the point, and certainly proves nothing against the *unity* of the soul of man. As the brain contains the inhibitory centres for the lower impulses, a morbid condition of certain parts of the brain may naturally afford the animal part of man an opportunity of asserting itself, as we frequently see in cases of softening of the brain. But it is difficult to arrive at any general conclusions from *Dementia paralytica*, for the course of this disease varies greatly according to its symptoms, whether vasomotor, psychical, or motor. Cf. on this subject Krafft-Ebing, *Lehrbuch der Psychiatrie*, 7th ed., Stuttgart, 1903, p. 573, etc.; Obersteiner,

Funktionelle und organische Nervenkrankheiten, p. 83; Bessmer, *Grundlagen der Seelenstörungen*, Freiburg i. B., 1906, p. 55, etc.

The æsthetical and ethical feelings and connections are among the first to be affected in *Dementia paralytica*, but this may be a result of a previous morbid change in the physiological centres of the sense feelings and impulses. If the centres are weakened, an undue ascendency is given to the peripheral stimuli. The weakening of the memory and of the power of attention (both being functions of the nervous system) is easily explained without assuming—as Juliusburger seems to do—that the higher entity of the soul is in a morbid state in softening of the brain. Juliusburger's allusion to melancholia has nothing to do with our present subject, for the first and most decisive symptom of this condition is a morbid state of the feelings. These are due to the body, and so melancholia is no evidence of the existence of disease in the higher intelligence. The action of the understanding is only impeded or more or less checked, because the patient's attention is turned exclusively to gloomy ideas. In this lies the psychical depression which characterises melancholia, and it is *primarily organic* because of its original connection with the feelings. Krafft-Ebing thinks (p. 229) that its physiological cause is to be sought in disturbances in the *nutrition* of the brain.

5. 'I will refer shortly to the remarkable condition which has been called "dual consciousness." These conditions can occur quite suddenly; whilst they prevail, people perform complicated actions, take long journeys, find themselves suddenly in strange places, and do not know how they came thither. A young man who was employed in Australia found himself suddenly in Zürich; he returned to his first personality through reading an advertisement in a newspaper, which reported his disappearance from Australia. In these cases it is not a subordinate mental organ which is affected, but a man's *whole personality* is completely changed.'

Answer to No. 5. The far-reaching disturbances of consciousness, which Juliusburger here denotes by the expression 'dual consciousness,' may be explained on the analogy of the conditions prevailing during *dreams or a state of dim-perception*, when the power is lost of judging of time or space, or of one's own social position,—dual consciousness is a continuous dream life. In the first place there is an absence of sense attention, which is so completely diverted from external sense activity by the force of certain subjective imaginary representations, as to destroy the power of recognising realities.

Many instances of disturbance of the sensory spheres are connected with a persistent state of dreaming. (Cf. Bessmer, *Grundlagen der Seelenstörungen*, pp. 54-55.) The *personality* of the patient is not altered in this state, as Juliusburger maintains—*he is only unable to recognise his personality.*

6. '*All these arguments prove irresistibly that there can be no simple soul unit of supernatural origin. Such an immortal simple unit could not suffer disease and decay.* The fact that disease can attack *the inner personality* of a man shows that we are not justified in asserting that disease attacks not the soul, but only its organ.'

Answer to No. 6. *All these arguments by no means prove that a simple soul unit cannot exist.* That it can exist is sufficiently proved by my critical answers to Juliusburger's five preceding remarks.

Juliusburger starts here, as before, from an obviously erroneous hypothesis. He represents the human soul as *purely spiritual*, with no intimate connection with the body, and in no part of its activity referable to any organs of the body. That such a soul could not suffer from disease, *i.e.* could not be disturbed in its action by any disease affecting the body, is self-evident and did not require such a long proof.

But according to Christian psychology, which Juliusburger was trying to disprove, the human soul is not purely spiritual, but *is a spirit united with a human body so as to form one complete whole.* The soul animates the body and is its essential form. Body and soul are but *one principle of activity* in the acts of perception, imagination, and sensation. If therefore the brain is diseased, the activities of the soul are affected—the lower ones directly, the higher ones indirectly, for the former are the necessary previous conditions of the latter.

My remarks up to this point ought to suffice to answer Juliusburger's arguments against the existence of a simple soul in man. The following two comments are of no importance as tending to establish his views.

7. 'The relations between bodily and mental life are intelligible only in one of two ways. Either we must accept Forel's theory of identity, that what, externally (!) regarded, appears as brain, is experienced internally—and constitutes the activity of the soul—and *vice versa*—what is effected by the activity of the soul, appears externally as brain. Or we must put ourselves at the point of view of the *monistic transformism,* and regard our mental

acts as an accumulation of psychological energy which cannot be distinguished from all other forms of energy. Thus our mental energy is only the transformation of the general energy which fills the universe.'

Answer to No. 7. Juliusburger is wrong in asserting that the relation between body and soul is intelligible only in the light of *the monistic theory of identity* or of that of *the monistic theory of transformation.* For (*a*) besides these two theories *there is at least one other, viz. that of the essential difference between the soul and body of man, of their connection to form one whole, and of their reciprocal action upon each other.* This is precisely the *dualistic theory,* which Juliusburger did not succeed in overthrowing in the previous six points of his discourse, and which therefore I still maintain to be true.

(*b*) The *theory of identity,* to which he referred, ought not to be known by the name of Forel, but by that of Fechner, who first evolved it. It is what has been called '*the stupid two-sided theory,*' which designates as 'mind' the inner side of the brain, and as 'brain' the outer side of the mind, *thus affording no scientific explanation at all.* Cf. on this subject C. Stumpf, *Leib und Seele* (Body and Soul), *Opening Address at the International Psychological Congress in* 1896, printed at Leipzig in 1903; also Wasmann, *Die monistische Identitätstheorie und die vergleichende Psychologie* (The Monistic Theory of Identity and Comparative Psychology), published in the *Biolog. Zentralblatt*, 1903, Nos. 16, 17. See also the remarks in my closing speech.

(*c*) The *monistic theory of transformation*, which seeks to explain psychical energy as being merely a transformation of 'the general energy which fills the universe,' is supported by no real proof, and is as contradictory as the 'Two-sided Theory' just mentioned. That mechanical energy can ever be transformed into physical energy is just as untrue to facts, and just as inconceivable logically, as that the *movement of atoms* can ever be transformed into *a thought*. Both forms of energy are different, though they are *in reciprocal action* upon one another, and the dualistic theory is certainly in a better position to explain this action than is the monistic theory.

8. 'As a specialist in mental diseases, I have a particular reason for rejecting the spiritual doctrine of the unity of the soul, because just at the time when alienists still accepted this doctrine, the insane were most brutally treated. As we became more monistic in our views, we treated the insane better, and the more monistic we all become, the better will our patients fare.'

Answer to No. 8. The advance in the treatment of the insane has nothing to do with monism, but with the modern growth of humane ideas, which are an inheritance from Christianity. Juliusburger would have expressed himself more correctly, had he said: '*The more Christian* we all become, the better will our patients fare.' The Christian commandment enjoining love of our neighbour has certainly produced nobler fruits in the form of *Christian charity* than the new monism has done. Men so distinguished in their treatment of the insane as Schroeder van der Kolk, Griesinger, and others, would decidedly resent being classed as monists because they initiated the reform in the management of madhouses. The monistic director of a famous asylum in Switzerland told me, sixteen years ago, that he preferred to take his attendants from the *Catholic*, rather than from the Protestant, cantons. Can we possibly imagine that the Catholic attendants held more advanced *monistic views*, and not rather that they possessed *more Christian* sense of duty ?

'I must shortly allude to another subject, viz. to our out-of-date ways of regarding and treating criminals. The futility of our struggle against crime is due, among other things, to the still prevailing doctrine of the unity of the soul. It is the outcome of this spiritualistic Psychology with freedom of the will as its central dogma.'

> This statement is incorrect. The futility of the struggle against crime is not due to the Christian doctrines of the unity of the soul and of the freedom of the will, but, on *the contrary*, to the contempt of these doctrines—a contempt which is spreading among the lower classes. If a man regards himself as nothing more than a superior kind of animal, and is convinced that he *inevitably* must yield to his impulses, however low they may be, he can certainly find a convenient excuse for every kind of crime. Monistic Ethics are therefore necessarily a cause of crime, not a remedy for it.

'Therefore for reasons of knowledge we must take up our stand upon a strictly monistic psychology. Speaking as a monist, I emphasise the identity and the connection of all beings. Written on the face of every human being, we seem able to read the inspiring words: "See, this is thyself." The recognition of this is, for us monists, the source of brotherly activity, but an integral part of monism is the theory of evolution. Man is becoming more and more conscious of evolution, and monism demands of him that he should consciously take part in it. To give free play to evolution is the monists' moral law, and the monistic doctrine of immortality is firmly rooted upon the necessity of labouring in this field of work, for the benefit of our contemporaries and of our posterity.'

Dr. Juliusburger does not seem to me to be quite consistent here. Darwinism and monism, *if applied logically* to man, would inevitably lead to the extermination of the 'unfit,' with all their possibilities of action, as being of less value and prejudicial to the successful breeding of a future race. This extermination would be 'giving free play to evolution,' but we must compare with it once more the noble law of *Christian love towards one's neighbour*.

IX. Dr. Plötz's Speech.

This speaker expressed his intention of discussing *the descent of man from beasts*. Father Wasmann had, he said, brought forward two arguments against it—the first being that *man possesses a special kind of soul, peculiar to himself*. This argument having been already refuted by another speaker,[1] Dr. Plötz did not think it necessary to do more than allude to it, but dealt more fully with Father Wasmann's second point, viz. that we must assume *a special creation* in the case of man. This

[1] Cf. Dr. Juliusburger's speech and my remarks upon it.

idea of creation was constantly recurring, and the speaker believed that Professor Plate had not sufficiently insisted upon the fact, that Father Wasmann's reasons for upholding the theory of creation were to some extent *logical*, and it was from the logical point of view that he himself wished to examine them.

Father Wasmann maintained that it was impossible to imagine the creation of the world, and the origin of organic life and of the human race as the result of a kind of natural evolution, because everything was so wonderfully adapted to the purpose which it was to fulfil, and was so marvellously highly organised.

> Dr. Plötz cannot quite have followed my reasoning if he sums it up thus. I showed in my second lecture that Creation was the *logical hypothesis* of an evolution tending to work out some design, but that Creation and Evolution were not *antagonistic* to one another.

Dr. Plötz went on to argue that my conception necessitated the existence of a *creator*, whose intelligence was so far superior to the world as to render the act of creation possible. Thus 'there is a sort of *logical motive* for Father Wasmann's assumption that a creator exists. *But if we once admit this, we must logically proceed to raise this further question:—As the creator is an organism so far superior to the world that the creation of the world can originate in him, we are forced to assume that he in his turn must have been created.*[1] *This leads us on to the creation of this creator, and so we may go on for ever.*'

In my second lecture I explained that *only an infinitely perfect Being could contain in himself the reason of his existence, and consequently be eternal.* Matter with its many limitations has not in itself even the reason for its *motion*, far less that for its *existence*, and so I argued that it must have been *created by God*, who alone has in Himself the reason for His existence. Nevertheless Dr. Plötz asks quite seriously in the midst of the discussion, '*Who created the Creator?*'

I wish to draw attention to the fact that this argument was put forward by a representative of modern German science, the editor of the *Archiv für Rassen- und Gesellschaftsbiologie*, and that it was uttered in the metropolis of German culture, at a scientific discussion held in the presence of two thousand people belonging to the 'Nation of Thinkers.' This may well give us ground for reflection!

As a sequel to my Berlin lectures, the *Vossische Zeitung* (No. 129, Sunday supplement,

[1] The reader must notice that Dr. Plötz calls the creator *an organism*, an anthropomorphic expression resting upon an absolute ignorance of the theistic conception of God. For the correct statement of the theistic view see Lecture II. p. 26, etc.

No. 11, March 17th, 1907) printed an article by Dr. R. Salinger, entitled *Monismus und Dualismus in erkenntnis-theoretischer Beleuchtung* (A Critical and Theoretical Examination of Monism and Dualism). The writer expresses his regret that practically no reference was made to philosophy during the discussion of my lectures. In order to supply the omission, and at the same time to overthrow Wasmann's reasoning in support of a Creator, Dr. Salinger presents his readers with the following 'little anecdote':—'A little girl of about six years of age was taking a walk one evening in the country with her mother. The sun was just setting, and was lighting up all the marvels of nature with its crimson and purple rays. Like other children who are eager for knowledge, the intelligent little girl was asking questions about the flowers in the fields, the clouds in the sky, and the animals in the forest; and, above all, she was anxious to know who had made everything that she saw. "Almighty God made it," was the invariable answer given by the mother to the child's questions. For a time this stereotyped formula satisfied the child, but then she asked with naïve simplicity: "And who made Almighty God?" It would seem that in asking this question the little girl showed a more genuinely philosophic spirit than Father Wasmann and his learned scientific opponents display.' Dr. Salinger has certainly not been fair to Dr. Plötz, for the latter brought forward exactly the same argument against assuming the existence of a creator as Dr. Salinger's 'intelligent little girl.' His philosophy therefore is just as superior to that of Father Wasmann and his other opponents, as is the philosophy of a six-year old child, who forms too limited a conception of the Creator, and so asks who created Him.

The speaker said that he believed the necessity of assuming an act of creation had prevented Father Wasmann from appreciating, at their just value, the facts which tend to show that man is descended from beasts.

He had absolutely denied the existence of *a missing link* between man and beast, and yet it was really the ape-man, which he thought to set aside with a mere wave of the hand and even with a joke.

Dr. Plötz was referring to the picture of the *Pithecanthropus erectus* as waiter, which I borrowed from the bill of fare at the International Zoological Congress at Leiden, and displayed for the amusement of my audience during my third lecture. If I had said that *scientific* men assumed the missing link to have presented this appearance, there would have been some justification for Dr. Plötz's indignation, but I was careful to guard myself against any possible misinterpretation.

Dr. Plotz said that he did not consider the matter to be by any means so simple. He referred to the five cranial lines on a photograph shown during the third lecture. (He was alluding to Macnamara's five cranial lines.) Father Wasmann had shown in a very convincing way that the cranial line of the Neandertal man and that of the Australian black almost coincided with one another. But it was possible for the audience to notice in silence that the cranial line of the *Pithecanthropus erectus*, the ape-man from Java, occupied a position midway between that of the Neandertal man and the outline sketched beneath it, which represented the skull of a gibbon. (The speaker meant to say a chimpanzee.) This was an opportunity of supplying Father Wasmann's omission.

From the size of the Java skull nothing can be inferred, as I said in my lecture, than that its owner must have been *a very large ape.*

To enable us to appreciate justly the relative positions of the Neandertal man and of the Pithecanthropus, Dr. Plötz regarded it as essential to compare their estimated or calculated *cranial capacity* with that of an ape on the one hand, and that of a man on the other. This comparison would justify us in speaking of a *missing link*; for the highest apes known to us, viz. chimpanzees and gorillas, have a cranial capacity of from 500-600 c.cm., the *Pithecanthropus erectus* has a cranial capacity of 900-950 c.cm., the Neandertal man about 1200 c.cm., and man at the present time from 1500 upwards. We might, said Dr. Plötz, congratulate ourselves upon having not one but two *missing links*, for the Pithecanthropus and the Neandertal man mark off approximately equal distances in the line connecting gibbons (*i.e.* chimpanzees) and men.

Dr. Plötz's arguments are based upon false premisses. His statements regarding the cranial capacity of a man do not agree with facts. According to Ranke, among the rural population of a department of Bavaria, the minimum cranial capacity is 1100 cubic centimetres, the maximum 1780; the average for a man is 1503, and for a woman 1335 (Ranke, *Der Mensch*, i., 2nd ed., p. 409). If, then, the Neandertal man had a cranial capacity of about 1200 cubic centimetres, he had more than the minimum at the present time, and so, from this point of view, he was a genuine man, and not a missing link between man and apes. However, Plötz seems to have given too high a number, as other scientists estimate his cranial capacity as amounting to only 800 c.cm.

The speaker wished to show that the *cranial capacities* supplied us with the best means of determining the position of this man-ape of the tertiary period, and of the Neandertal man. Since man has walked upright—and the Pithecanthropus must have walked upright because of the shape of his femur[1]—natural breeding has tended to a development of the brain, rather than to perfecting the rest of the body. As a matter of fact, at the present time, we see that the highest and the lowest races among men are distinguished chiefly by the difference in their *cranial capacity.*

Dr. Plötz's theory is not supported by facts. Whereas among Europeans the maximum cranial capacity is under 2000 cubic cm.—the highest number given by Welcker is 1870—Virchow found that the capacity of an average male skull in New Britain was 2010 cubic cm. The smallest known capacity is that of a female Wedda skull in Ceylon, which measured only 960 cubic cm. (Ranke, *Der Mensch*, i., 2nd ed., p. 409). This approaches very closely that of the Pithecanthropus, even if we accept 800 cubic cm. as being the correct number; but it must be borne in mind that the Weddas are *dwarfs*, and the Pithecanthropus was *a giant ape*, and so Dr. Plötz's whole argu-

[1] Whether this is really the case requires examination with Röntgen-rays, by means of Walkhoff's method. Until this has been done, it is not certain that he walked upright, although it is probable.

ment is falsified. The *absolute* cranial capacity does not tell us whether the owner of the skull had a large or a small brain *in proportion to his size.*

According to Dr. Plötz, increased cerebral activity tends to adapt man to his surroundings, and also his surroundings to him, much better than the greater development of the extremities that he once possessed; therefore the consideration of the form of the body is not so important as the consideration of the cerebral capacity, because from this we can trace the degree of increase in the size of the brain that filled the cranial cavity. As the organisation of the brain advanced, the intelligence of man developed.

There is an element of truth in these assertions, for the difference in *the formation of the brain is the chief distinguishing mark* between man and beast, as far as their bodies are concerned. I discussed this point in more detail in my third lecture (p. 56, etc.). But as the development of the human brain is more perfect than that of the highest apes, because the human brain is the instrument of intellectual activity, I added the remark that all the bodily differences between man and beast are ultimately a result, a function, of their mental difference.

Dr. Plötz makes a mistake, however, in

laying such undue stress upon the *cranial capacity.* It can only supply us with information as to the size of the brain, but it tells us nothing about the far more important differences of the cranial formation; still less does it enable us to judge of the intelligence of the owner of the skull. Comparative anthropology has declared the cranial capacity to be of no use as an absolute test in determining the stage of intellectual development. This is plain from the following table (Ranke, *Der Mensch,* ii. p. 482), which gives the average cranial capacity of some modern and some ancient races of men.

	Cubic cm.
Parisian of the twelfth century, . .	1532
Modern Parisian, ...	1558
Modern inhabitant of Lower Brittany, .	1560
Prehistoric northern cromlech-builder,	1580
Spanish Basque, ...	1584
Gaul, ...	1585
Prehistoric cave-dweller in Cro-Magnon,	1590 (1640)
Modern inhabitant of Auvergne, . .	1598
Prehistoric skull from the Homme mort cave, ...	1606
Skull from the prehistoric outpost at Solutré, ...	1615

These figures speak plainly enough. Mankind must have been steadily becoming more stupid from the Alluvial Period to the present day, if Dr. Plötz were right.

In conclusion, Dr. Plötz remarked that the

oldest existing skulls, which were near that of the Neandertal man in age—such as the Krapina skull, etc.—were all characterised by having pronounced superciliary ridges, which scarcely occur at all among the races of men now existing. Also the lower jaw is almost devoid of chin. *The chin is very characteristic of mankind,* more so perhaps than any other external feature. Dr. Plötz argued from these facts that, from the anatomical point of view, we were not justified in refusing to accept the Pithecanthropus as a missing link.

> I answered these remarks in my closing address. The very early race that inhabited the Neandertal, Krapina, etc., cannot be regarded as a *missing link* between apes and man, because in all essential points they represent a genuine human type. This race had no particularly close connection with the Pithecanthropus, as the latter belonged to a branch of the family of apes. (See my third lecture, p. 74.)

The proceedings were interrupted at this stage by the interval of five minutes to which I have referred in my preliminary remarks on the evening discussion, p. 89.

X. Dr. Schmidt-Jena's Speech.

The speaker introduced himself to his audience as having been for *many years Professor Ernst Haeckel's assistant* at Jena, and as now being the General Secretary of the German Monistic Association; he thought he had therefore an excuse for making a few remarks upon Father Wasmann's lectures. In the first place he wished to say a few words on Haeckel's views. There were three chief points which he wished to emphasise: Firstly, he challenged Wasmann's assertion that Haeckel had *for forty years consistently confused the doctrine of evolution with Darwinism or the theory of selection.* This statement was, according to Dr. Schmidt-Jena, absolutely false, and the contrary was the truth.

In my closing address I showed that Haeckel, as a popular speaker and writer, often expressed himself quite otherwise than he has done in his more important works, where he speaks as a specialist in zoology. I may add here some further proofs of what I have just said. Compare, for instance, the lecture delivered at Eisenach on the *View of Nature taken by Darwin, Goethe, and Lamarck* (Jena, 1882), in which no one could fail to perceive that he identifies Darwinism with the doctrine of evolution, and does so for a specific purpose.

On page 6 of his *Welträtsel* (Bonn, 1899), in speaking of the relation in which Darwinism stands to the doctrine of evolution, Haeckel says: 'It will never be forgotten that the merit of having experimentally laid the foundation of this highest (!) philosophical conception (of evolution) must be ascribed to the great English scientist Charles Darwin; in 1859 he established on a firm basis that theory of descent, which the genial French naturalist, Lamarck, recognised in its general outlines as long ago as 1809, etc.' It can hardly be maintained that we have in these words a clear distinction between Darwinism and the doctrine of evolution.

Dr. Schmidt-Jena, however, maintained that in his book on *General Morphology*, written forty years ago, Haeckel had drawn a very sharp distinguishing line between, on the one hand, the theory of evolution in general, and the theory of organic evolution in particular (which he called Lamarckism, after the French zoologist Lamarck, who was its real author) and, on the other hand, Darwin's theory of selection, which he designated Darwinism. If in later years he used the name Darwinism in a wider sense, he hardly ever forgot to add, by way of explanation, that he meant the whole theory of evolution, and especially that of organic evolution, and where he did not definitely state this, it was perfectly clear from the context that such was his meaning.

The speaker here acknowledges what he had previously denied, viz. that Haeckel very often used the word Darwinism in a *misleading way*, as synonymous with *the whole theory of organic evolution*. This concession on the part of one who was for years Haeckel's assistant is certainly very important. As further illustrations of his confused use of the name Darwinism, I may refer to some passages in Haeckel's Berlin lectures, delivered in 1905, entitled *Der Kampf um den Entwicklungsgedanken* (The Struggle regarding the Idea of Evolution). On p. 20 he distinguishes clearly between Darwinism and the theory of evolution, and states definitely that by Darwinism we ought, strictly speaking, to mean only Darwin's theory of selection. But a few pages further on he forgets this distinction and confuses Darwinism with evolution in his old fashion. On p. 32 he declares that I had explained my observations on the inquilines among ants 'quite in the Darwinian sense.' On p. 34, in speaking of an alleged 'concession to the Church' on my part, he says that the organisms developed 'in accordance with the Darwinian laws.' On p. 75 he calls me the 'Darwinian Jesuit,' and so on. In his use of Darwin's name,

Haeckel certainly deserves the imputation of double dealing or of 'faking his accounts.'

The second topic with which Dr. Schmidt-Jena wished to deal concerned Haeckel's pedigrees. He remarked that on the previous evening (*i.e.* at my third lecture, see pp. 79, 80) Father Wasmann had referred to one such pedigree, and had remarked that *comments were superfluous.* The speaker begged to be allowed, nevertheless, to comment on it. When Haeckel compiled his first pedigrees between 1860 and 1870, he was told that they were fantastic, and yet it was assumed that he intended them to be taken literally. In the second edition of his *Natürliche Schöpfungsgeschichte,* he wrote: 'On this occasion, as well as with reference to my other hypotheses regarding evolution, I protest against having any dogmatic significance ascribed to them. They are merely first attempts.'

This sentence stands in every edition of the *Natürliche Schöpfungsgeschichte,* from the second to the tenth; and in his other works, and especially in his great book, *Systematische Phylogenie* (the race history of living organisms), Haeckel has again and again tried to protect himself against this dogmatic interpretation of his pedigrees, and has repeatedly insisted upon their purely hypothetical character. But these assertions on his part availed nothing; even at the present day he is charged with having himself taken a dogmatic view of his pedigrees.

In spite of this statement, the truth remains that, in virtue of his double character, Haeckel often made very different assertions in his *popular* works from those in his *scientific* writings. Cf. the striking passage, quoted in my closing address, on the subject of Haeckel's pedigree of primates (apes and men).

On the same subject Haeckel says in his *Welträtsel*, p. 99: 'In the last twenty years a considerable number of well-preserved fossil skeletons of anthropoid and other apes have been discovered, and amongst them are all the important intermediate forms, *which constitute a series of ancestors connecting the oldest anthropoid ape with man.*' If any one can regard a statement of this kind as a modest hypothesis, and not as an apodictical, dogmatic assertion involving the constitution of pedigrees, that person must have a very peculiar idea of what an hypothesis is.

The third point to which the speaker referred was the *biogenetic principle.* He said that Father Wasmann had stated this law—the fundamental law of organic evolution—as Haeckel called it, in the following way:—'Ontogenesis, or the evolution of the individual, is the repetition in brief of the phylogenesis, or the evolution of the race.' He had asserted, moreover, that according to this law the ontogenesis must reproduce the phylogenesis *in detail.*

This statement of Dr. Schmidt-Jena's is not accurate. When expounding the biogenetic principle (see p. 58, etc.), I said nothing at all about an exact repetition of the phylogenesis in the ontogenesis. These words were only used later (p. 70) as an *argumentum ex absurdo* against the descent of man from apes. But on this subject I may refer to my closing speech. It would be interesting to learn from Dr. Schmidt-Jena, by what theory of Cænogenesis, or 'subsequent interference with evolution owing to adaptation,' he would account for the fact that young apes bear a much more marked facial resemblance to man than old apes do. This resemblance cannot possibly be due to subsequent adaptation, therefore it forms a good *argumentum ad hominem* against the direct descent of man from apes.

Dr. Schmidt-Jena explained further that Haeckel had made a very important addition to his law, viz. that *the repetition of the history of the race in the evolution of the individual is affected by the action of heredity and adaptation.* He maintained that Haeckel had shown that, strictly speaking, no phylogenetic stage could be reproduced in the ontogenesis, because inevitably adaptation to special circumstances gives the evolution of a germ another tendency than we should expect, if we were only to look for phylogenetic reproduction in accordance with heredity.

He pointed out that Haeckel had used two distinct names to designate these two aspects of the biogenetic principle, viz. *Palingenesis*, or original reproduction, and *Cænogenesis*, or subsequent modification. He admitted that Haeckel's theory had been subjected to much criticism, and that his law had been found not to correspond with ascertained facts, inasmuch as the changes undergone by embryos and by young animals were too great to be accounted for thus, and could not be regarded as a reproduction of the evolution of the race. Nevertheless, Father Wasmann himself had recognised the importance of the law in several respects, and had adduced the best instances of its applicability.

The speaker made a mistake here. I never recognised *the biogenetic principle as such*, either in my third lecture or in my book on *Biology and the Theory of Evolution*. The instances adduced by me, to which Dr. Schmidt-Jena referred, were exceptional cases *of relatively rare occurrence*, in which the development of the individual gives us a clue to the evolution of the species (see p. 62). But the fact that these cases are exceptional and of rare occurrence shows that the biogenetic principle is not a *general law*. The reader is invited to compare the remarks in my closing address on the subject of this biogenetic principle.

Dr. Schmidt-Jena claimed to have proved in a little work published some years ago, that the biogenetic principle was applicable to *the development of every individual*, precisely because it keeps in view both heredity and adaptation. He thought that in the same work he had shown another eminent scientist to have been wrong in assigning another interpretation to this principle.

This other eminent scientist was no less important a person than Professor Oskar Hertwig, director of the Berlin Institution for Biology and Anatomy. No one can seriously think that Hertwig's opinions were refuted in the insignificant pamphlet written by Haeckel's assistant. See Hertwig's more recent work 'Das biogenetische Grundgesetz nach dem heutigen Stande der Biologie' (The Biogenetic Principle considered in the Light of Modern Research in Biology) (*Internationale Wochenschrift für Wissenschaft und Technik*, 1907, No. 2, etc.). In my third lecture I used Hertwig's arguments against the biogenetic principle. If his 'other interpretation,' or rather refutation, of the principle is a failure, Dr. Schmidt-Jena would have done well to prove it, and not to expect us to accept his bare assertion.

As to the relation in which the doctrine of evolution stands to Theism and Monism, Dr. Schmidt-Jena said he only wished to remark that, in his opinion, evolution, as such, was neither theistic nor monistic. Father Wasmann laid stress upon theism, and made it the foundation of the doctrine of evolution and of all other sciences, whereas to a monist the monistic philosophy of nature was a deduction from the doctrine of evolution, at which a scientist could arrive, if he chose, but he was not bound to do so.

The General Secretary of the Monistic Association certainly made a very true and very important concession, when he acknowledged that the doctrine of evolution as such was neither theistic nor monistic. In numerous passages in his works, Haeckel has always represented the doctrine of evolution as being an integral constituent of the monistic view of life. Cf. *Der Monismus als Band zwischen Religion und Wissenschaft*, Jena, 1893, p. 19, etc.; and *Die Welträtsel*, Bonn, 1899, pp. 271, 383, 437, etc. The speaker was, however, inaccurate in his representation of my opinions. I consider the theistic view of life also to be a *deduction*, strictly conformable to reason, from the scientific doctrine of evolution. *If* evolution has taken place at all, it must have had its origin in the creative action of God, who formed a world capable of evolution. Considered in its logical aspect, this is a deduction, but it can also be regarded as a fundamental principle in its relation to actual facts.

In conclusion Dr. Schmidt-Jena wished to say a few words on the frequently mentioned metaphor of the *rock of Christianity.* In the course of the discussion it had been proved (!) that the waves of science had already washed away a good deal of it; first the Ptolemaic System, then Permanence of Species, then belief in miracles (which is now replaced by the theory of conformity to law)—all these were integral parts once of Christianity, and have now been given up, because it was vain to attempt to resist the progress of science.

What Dr. Schmidt-Jena said about the integral parts of Christianity does not call for a long answer, for neither the Ptolemaic System nor Permanence of Species, nor any belief in miracles, which could conflict with the acceptance of natural laws, ever formed integral constituents of Christianity.

Finally, Dr. Schmidt-Jena said that *even in the case of Father Wasmann, theism was shaken, and was in danger of being broken down altogether by the advancing tide of the doctrine of evolution.* He supported this interesting assertion by the following evidence:—Theism regards as necessary *an incessant interference* on the part of a personal God in the course of events in the world; but Father Wasmann declares this interference to be necessary only on two or three occasions, and possibly only on one, viz. *when the original miracle of creation was wrought.* As a matter of principle it was unimportant whether we admitted a miracle—a violation of the laws of nature—to have taken place once or many times; a true scientist could not regard such a violation of law as being ever possible. But after Father Wasmann had abandoned this theistic theory of the incessant interference on the part of God in the course of events in the world, he had arrived at another theory, which might be called deistic, according to which the divine Being constructed the machinery of the universe, gave it its laws, set it in motion, and then let it work without further interference. This deistic conception bordered on pantheism, and therefore, in the speaker's opinion, *Father Wasmann needed but to take one short step further and he would arrive at pantheism, which every one knew to be Haeckel's view.*

These philosophical statements regarding theism, etc., seem to be derived from Haeckel's *Welträtsel*, Parts III. and IV. Any textbook of the Christian theodicy would have made it clear to Dr. Schmidt-Jena that theism does not assume any *arbitrary interference on God's part* in the course of events in the universe. He was wrong, too, in calling creation a miracle, for before it there were no natural laws, and consequently such laws could not have been violated by creation.

Dr. Schmidt-Jena is plainly confused as to the differences between Theism, Deism, and Pantheism, therefore I will here give a short and accurate statement of the distinctions between them.

(*a*) According to Theism, God is the *infinitely perfect and eternal Being*, and so God's nature *differs essentially* from the nature of the world, which consists of finite and imperfect things. The finite could proceed from the infinite only by an *act of creation* on God's part. In virtue of His infinity, God is most *intimately present* in all creatures, and because of *the continuous dependence* of the finite upon the infinite, He is *active* in all creatures through their preservation (*conservatio*), and He participates in all their actions through co-operation (*concursus divinus*).

He does not *arbitrarily interfere* with the working *of the natural laws* which He Himself has laid down, because to do so would be incompatible with His infinite wisdom, which is identical with His power. Therefore in the natural order there are no miracles, *i.e.* exceptions to the natural laws. But over and above the natural order, God has given a supernatural order to those of His creatures who are endowed with reason, and He has done this through the Christian revelation. Miracles occur only to further the aims of this supernatural order.

(*b*) Deism denies altogether *the existence of the supernatural order*, and recognises only the natural order, without regarding the *conservatio* and the *concursus divinus* as necessary. Deism has nothing in common with Theism except the fact, that both regard God as essentially different from the world.

(*c*) Pantheism denies the essential difference between God and the world, and asserts their *substantial identity*. It borrows from theism the ideas of the universal presence of God and of His co-operation in the actions of all creatures. There are many varieties of pantheism. At one moment it regards God as of primary importance, and the universe as secondary manifestations of Him ; at another moment pantheism becomes *mere naturalism*, ascribing reality only to the universe, but keeping the conception of God as a kind of synonym for ' universal nature,' in order to conceal the atheism which really underlies it. *This last is the pantheistic conception of the universe which Haeckel upholds.*

Having thus explained these distinctions to Dr. Schmidt-Jena, I should like to ask him, as a favour, not in future to confuse my theistic opinions with deism, and still less not to connect them with Haeckel's atheism—to do so would be an absolute misrepresentation based on falsehood.

In the course of the speeches delivered in the spring of 1907, when he was travelling about in his capacity as General Secretary of the Monistic Association, and especially at Vienna, where he tried to gain adherents to that Association, Dr. Schmidt-Jena asserted several times that Father Wasmann had already passed from theism to deism, and was on the verge of becoming a pantheist. If the German Monistic Association is forced to have recourse to such means as these to win adherents, it certainly does not make for 'enlightenment.' I fully concur with the sharp criticism pronounced by Professor Reinke at Herrenhaus on May 10th, 1907, when, in speaking of the exertions of this Monistic Association, he declared them to be a common danger to German culture.[1]

In his lecture on Natural Science and Religion (*Die Propyläen*, March 13th, 1907, No. 24) Professor Reinke propounded what are practically the same opinions regarding the relations existing between scientific research and

[1] See also Reinke's article on Haeckel's *Monism and its Supporters: a Free Word for Free Science.* Leipzig, 1907.

the theistic theory, as are contained in my Berlin lectures.

XI. Dr. Thesing's Speech.

The speaker began by saying that his audience must be as tired of the discussion as he was himself, therefore he would make his remarks as short as possible, especially as the most essential scientific points, which he might otherwise have discussed, had already been dealt with by one or other of the previous speakers. He was surprised to see how completely they were all of one mind in their opposition to Father Wasmann, and he was glad to have an opportunity of expressing his own agreement with him on several points, and particularly on some very important matters, *e.g.* on the view which Father Wasmann took of Darwin's theory of selection, to which it was wrong to ascribe as much importance as many people were still inclined to do. It might be regarded as *a subsidiary hypothesis*, presupposing the presence in the organism of something else, viz. *of internal forces.* Dr. Thesing declared this to be his own opinion, and he said that he differed from Father Wasmann only in one 'small detail'; whereas the latter regarded *the interior causes as a part of the divine Will or of the Deity,* and so proclaimed them to be something *inexplicable*,[1] he himself maintained that, in the case of these interior forces also, the demand for their quantitative determination must hold good. The word Vitalism was in bad odour in scientific circles, but, nevertheless, he would to some extent acknowledge himself to be a vitalist. Those interior forces might be designated by any name—we might call them physical energy, if we liked; but for this form of energy, as for all other forms, the demand held good, that *it should admit of quantitative determination.*

If Dr. Thesing meant that, in order to be scientifically explicable, a thing must admit of *quantitative determination*, I cannot agree with him. There are in nature qualitative differences also, and we discover these in the course of our observation of *vital processes*, in as far as they are really such.

The growth of a tree may be measured quantitatively, it is true, but living growth is something differing in quality from a mere addition of new to old atoms. Further, the word 'Vitalism' is in bad odour, not among scientists, but among materialists. Dr. Thesing, in acknowledging himself to be a vitalist, does so in company with Driesch, Reinke, and other eminent biologists; nevertheless, the opinion

[1] I have no idea where I am supposed to have said that the interior laws of evolution affecting organic life were 'a part of the divine Will or of the Deity.' Cf. my second lecture, pp. 29, 35.

> which he stated is, as I have already remarked, not a matter of vitalism but of mechanics, for the phenomena accompanying psychical processes and the elements in the organic processes, in as far as both are mechanically measurable, do not possess the quality which distinguishes these phenomena as psychical or vital. Life *as such*, both psychical and organic, is something *not quantitatively* measurable, precisely *because it is not mechanical.*

The speaker agreed with Father Wasmann in thinking that his theory of a *polyphyletic* evolution could be defended from a purely scientific point of view. It was *absolutely impossible* to prove a monophyletic origin of all living creatures.

> As opposed to the 'supernatural species,' which Plate had ascribed to me in his speech, this remark of Dr. Thesing's was plainly very opportune.

There were a few points on which the speaker *did not agree* with Father Wasmann. The latter, he said, had asserted that matter had not existed from all eternity, and that he assumed a divine act of creation in order to account for its existence. Therefore, argued Dr. Thesing, God created matter, and God is eternal, and the question inevitably presents itself: 'What is God? Is He a point, a nothing, or what is He?' We can only say that, if we wish to connect anything at all with this idea of God, we can think of Him only as a God *whom we can imagine.*

There is here a great discrepancy between my philosophical opinions and those of Dr. Thesing. According to Christian philosophy, God is not a point, nor a nothing, nor a bodily form such as we can imagine by aid of our senses, but He is a *pure spirit, universally* present in virtue of His infinity. That we must *imagine* God, in order to be able to think of Him, is an anthropomorphic view that is quite untenable. Haeckel had such an idea of God, when he said that he could think of the personality of God only *in bodily form* as a 'gaseous vertebrate.' Dr. Plötz, too, had a similar idea, when he spoke of God as an '*organism*' (p. 177). This erroneous idea has spread unfortunately very widely in so-called educated circles, as a consequence of the publication of Haeckel's *Welträtsel* and similar books. People believe that the 'Personal God' of Christianity must be imagined as a sort of higher mammal, and as an illustration I may quote a letter written in Berlin, which seriously *propounds the following objection to the theistic conception of God* :—

'To imagine a personal (=corporeal ?) Creator as the first living being is probably impossible, for the question arises involuntarily: "Whence does this highly developed being suddenly come?" He must as such consist of an organic mass, composed of cells. But, to quote Virchow's saying, with which you probably concur, *omnis cellula ex cellula*, it is obvious that this being must have been evolved from some primitive cell. The assumption that the first being was a simple mass like a cell, is far more likely to be correct, and is more simple than your assumption that there was in the beginning *a highly organised Creator*.

'Hoping for a speedy answer, I am, etc.'

It is true that if we regard the personal Creator, the *ens a se* of Christianity, in this way, then Dr. Plötz's question: 'Who created the Creator?' is as apt as the question: 'Who laid the egg from which the Creator was hatched?' It is a lamentable characteristic of our age, that Haeckel's influence on philosophy has reduced men, even among the educated classes, to have recourse to such expedients. And our nation was once the nation of thinkers!

I need scarcely say that these remarks do not apply to Dr. Thesing, but were evoked by his question regarding the nature of God.

Dr. Thesing went on to say that a subject must have an object, and the conception of the person imagining a thing presupposed immediately that which he imagined. What else could this thing imagined be but matter? *Thus we might see, that from the assumption of the eternity of God followed immediately the eternity of matter.* A person who imagines nothing (and apart from matter nothing can be imagined) is nothing at all. Matter can only be defined as an aggregate of definite conformities to law. If, then, we allow that matter with its conformity to law has existed from all eternity, *we can very well dispense with its starting-point, i.e. with God.*

This argument contains the following four logical blunders, to which I need only draw attention shortly.

(1) In order to think of God, we must be able to imagine Him *by aid of our senses.* Professor Dahl rightly laid stress upon the inaccuracy of this idea.

(2) The idea of imagination presupposes the existence of the thing imagined. If this were true, no artist could ever produce a new work of art.

(3) God's knowledge is subject to as many limitations as man's, who cannot think of anything which he cannot imagine by aid of his senses.

(4) The conformity to law on the part of matter is something that we can imagine by aid of the senses. In reality it is something of which we can only think, underlying the pheno-

mena which we can imagine. I do not believe that such a line of argument suffices to render us able to dispense with God.

Dr. Thesing proceeded to say that Father Wasmann adhered to the Christian theory, according to which God was an absolutely perfect being. Thus from absolute perfection was derived the world, which was imperfect. He maintained this to be a contradiction very hard to comprehend.

If we, like the Monists, regard the finite as identical with what is divine and infinite, we certainly are involved in a totally inextricable difficulty ; but there is no contradiction in the theistic theory, which represents the finite as proceeding by way of creation out of the abundance of the infinite.

Dr. Thesing referred next to spontaneous generation. He said that Father Wasmann maintained rightly the impossibility of proving a spontaneous generation. We might show it to be probable that a living being arose out of what was inorganic, but *we could not prove it.* Was it, however, a necessary consequence that we must assume the existence of something apart from matter, and *a divine creation* ? Dr. Thesing did not think so. He said there were many other imaginary things which had just as much justification. While stating expressly that this was not his own opinion, but only an objection that might be raised, he asked why we were bound to assume that life ever had a beginning, why life could not just as well be eternal as matter? The answer to that question would be that, in accordance with the theory of Kant and Laplace, we know the world has been developed, and that it has gradually come into being from a condition of molten heat. Against this theory the speaker referred to the cosmozoic theory, which has found many supporters.

> I dealt with this objection in my closing speech (p. 213, etc.) and need not discuss it here.

Finally, Dr. Thesing insisted upon the fact that, in his opinion, it was a waste of time to try to disprove religion or the idea of God from the point of view of natural science. Even the dogmatic conception of God, upheld by Father Wasmann, could not *with certainty be proved to be impossible* by any one limiting himself to the domain of natural science. Religion and science occupied two totally distinct regions, and dealt with absolutely different problems, and they ought to be kept apart and not confused, as they had been on this occasion.

CLOSING SPEECH BY FATHER WASMANN, S.J., 11.30 P.M.[1]

LADIES AND GENTLEMEN,—

When it was first suggested to me to deliver these lectures, I expressed a wish to be permitted to speak at a discussion which should follow them. As to the special form which to-night's debate has taken, my thanks are due to Professor Plate for the proposals that he made with regard to it. My hopes that the discussion would be kept within the limits of a purely scientific treatment of our subject were to some extent realised in the case of Professor Plate himself, if I may except the conclusion of his speech, in which he pronounced me to be no true scientist. As to the other speakers it is impossible for me to express any such general opinion; I shall perhaps be able to comment upon one or two remarks made by them individually.

You will allow that it is very difficult for me to answer such a number of objections in a short time. I do not intend to speak for more than half an hour, for it is already close upon midnight. I must certainly answer Professor Plate first of all, as I

[1] In my closing speech I did not allude to the fact that the regulations for the debate, which I had accepted, had been altered without my knowledge and consent. (See the preliminary remarks on the evening discussion, p. 85.) At that time I was not aware who had formed the 'majority' that authorised this alteration. For details, see p. 87.

regard him as a prominent supporter of monism, and also as an eminent colleague in my own special department of zoology. I should like in the first place to emphasise one point. Although I have frequently insisted upon it in my lectures, Professor Plate does not distinguish with sufficient clearness between the *natural* and the *supernatural* orders. I did not come here to lecture you upon *theology*; I came with the intention of speaking about *the theory of evolution,* and of presenting to you the most essential points for consideration regarding the relations between it and Christianity. To discuss the matter fully is impossible in lectures of this kind before so mixed an audience, and this statement explains a great deal. You will permit me, however, to touch upon a few points in greater detail.

The remark that I have just made is aimed at Professor Plate's first observation, which was to the effect that my lectures and the present debate were to deal with *the struggle between the Church and natural science.* This assertion is absolutely false. I have not come here as a representative of the Catholic Church or of the Society of Jesus, but as Erich Wasmann, as a zoologist, whose own deepest convictions make him personally adopt the standpoint of Christianity. Thus *everything else which has been dragged into the discussion*—the Index (applause), the burning of Giordano Bruno, and other such things, *will be disregarded and passed over in my answer, simply because they do not belong to our subject.* They were only attacks upon me, made in a sectarian spirit.

Professor Plate and some others among my opponents maintain that I have a twofold character, that I am at once a theologian and a scientist. I am thankful for this twofold nature. The scientist and the theologian have only to practise self-control in one person, and this may be good for them both. (Laughter.) On higher matters the theologian speaks first, but it is an excellent thing for him to have the scientist at his elbow, to give him a little help now and then, and to put him on his guard against false opinions on scientific subjects; and it is good for the scientist, in his turn, if he has the theologian at hand, for a theologian is, as a rule, at the same time a fairly good philosopher, and *philosophy is absolutely indispensable to a scientist.* While my opponents have been speaking this evening, I have noticed again and again that I had been completely misunderstood on various points; this misunderstanding might have been averted by a more thorough training in philosophy. By such training, I mean that *strictly logical training* which forms an important part of our course of studies, but which is often neglected elsewhere.

The first 'scientific problem' mentioned in Professor Plate's objections concerns *the origin of life.* In my second lecture I remarked, that this was *not an ecclesiastical problem.* As soon as science can demonstrate to us that spontaneous generation actually occurs, and that it does not contradict the facts of biology, we shall willingly surrender the postulate, that some special action on the part of the Creator upon primitive matter was required for the origin of the first organisms. *This is only an extremely conditional postulate.* It cannot be said that in denying spontaneous generation I was influenced by ecclesiastical prejudice—on the contrary, I spoke as *a scientist*; and many other scientists, who care nothing for Christianity, have taken the same view, and have regarded the theory of spontaneous generation as one that could not be adopted *from the point of view of natural science.* It is clear that it cannot be adopted; but some one comes and says that it is a *philosophical postulate* for the scientist! This is a real contradiction. How can anything, which is at variance with scientific facts, be a postulate in the domain of philosophy? This is a contradiction in terms.

With regard to *the existence of matter* and to *the idea of creation*, a great deal has been said by Professor Plate and others, which shows clearly that the philosophical arguments in my second lecture were not understood by them.[1]

The lateness of the hour prevents me from

[1] My opponents have an opportunity of referring again to the lecture, now that it is printed.

dwelling upon the subject. I must, however, elucidate one point, viz. the statement that creation admits of no explanation.

An explanation of the first appearance of matter and of the first appearance of the laws governing it, is not possible, if we understand thereby *an explanation given by natural science* ; for this starts with the assumption that matter and its laws exist. But *in the philosophical sense* an explanation of creation is possible. Philosophy shows us plainly that *matter is finite* ; the conception of matter and its properties involves its being essentially limited and finite. It is therefore inherent in its nature that *it cannot of itself have existed from all eternity,* for this is possible only in the case *of a being of infinite perfection,* an *ens a se,* as ancient philosophy and theology worded it. This being we call *the personal Creator,* the being existing of Himself for all eternity, and *having the reason of His existence in Himself.* Precisely because He has the reason of His existence in Himself, He was able out of the abundance of His own infinite perfection to evoke the finite out of nothingness, and this is what we call *creation.* Creation was not necessary—it was a free act of God.[1]

[1] This can be deduced philosophically from the fact that the things in the world are in their nature finite and limited, therefore *they cannot be essentially necessary* ; they are, as ancient philosophy expressed it, *entia contingentia.* Whether an atom more or less exists in the universe is quite indifferent ; but if this is true of atoms singly, it must be true of them collectively. God alone, in virtue of His absolute existence, is *the one necessary being,* the *ens necessarium.* If God, by a voluntary exercise of His omnipotence, created the world, *this exercise of the divine Will* was

Thus *by means of philosophy* we can easily explain the idea of creation, in fact we can explain and understand it far more easily than that of the eternity of matter. Of this a *scientist* can only say: 'I do not know whether it ever had a beginning; I do not know whether it ever ceases to exist, because, as far as my scientific experience goes, there is no origin and no destruction of matter.' This is quite right; but if the scientist considers the question *philosophically*, he must, nevertheless, say: 'The ground of the eternity of matter, of its existence without any beginning, is not inherent in the conception of matter. In fact, these two conceptions, 'matter' and 'eternity,' are contradictory. Only a being not subject to change, and having in itself the reason of its existence, can be eternal. This is inconceivable in the case of matter, because it is imperfect and subject to change. Therefore we are forced to explain the origin of matter by means of creation—and this is where our *philosophical* ideas begin.'

Professor Plate came forward, moreover, as a champion of the hypothesis of *spontaneous generation*. He almost seemed to believe that I had said the first living creatures were produced by God by

not something unnecessary or contingent, for it is identical with God's being, because this is *absolutely simple*, and therefore in it there is no distinction between existence and activity, as there is in the exercise of the human will. The act of creation was *free* on God's part, because the world is not a possession *necessary* to God, and consequently its existence is not and cannot be willed by God as *a necessity*.

means of a creative act, in such a sense that the matter composing them was also newly created. This is a complete misunderstanding.[1]

I accept the theory that the first living creatures were produced from inorganic matter, in the sense that they were really composed *of inorganic substances.* But I cannot discover in inorganic matter any reason which could convert lifeless atoms into *the first living creatures.*

Perhaps I may be permitted at this point to deal with two subjects mentioned by the last speaker, Dr. Thesing, whose remarks were on the whole very much in accordance with my own views,—I mean the *Cosmozoic theory,* and the eternity of living matter. This hypothesis also is not tenable, quite apart from the fact that it leaves the origin of life unexplained. Many very pretty and ingenious speculations have been attempted by the supporters of this theory, Preyer, Thomson, Helmholtz, Richter and Arrhenius, etc. Some have imagined that the germs of life were brought to our earth by means of meteors or as cosmic dust. That cannot be, for the meteors must have been in a state of incandescence whilst passing through our atmosphere, and the cosmic dust, which was once alive, and is supposed to have retained its vitality, can be regarded only as a fiction.[2]

[1] Cf. p. 29 in my second lecture, where I expressed myself very clearly on this subject.

[2] In a book entitled *Das Werden der Welten* (Leipzig, 1907, chap. viii.), Svante Arrhenius has recently developed more fully his ingenious theory

Others again imagine that the matter existing on the earth was not originally separated into organic and inorganic, but was in a kind of intermediate state; but this appears to me to be physically impossible. The subject has been discussed more fully in my work on *Biology and the Theory of Evolution* (3rd ed., p. 197 *et seq.*, p. 208 *et seq.*), which Count Hoensbroech quoted to-day.

I now return to my first opponent.

He laid great stress upon the fact that *definite elements,* twelve in number, comprising five in albumen, constitute all living creatures. That is quite true, but the question is how, from these elements, the first living creature came into being, a creature *really alive*, having power to assimilate nourishment and to propagate others like itself. I agree on this point with Professor Oskar Hertwig, who in his *General Biology* makes the very clever and shrewd remark, that these theories are just repetitions of the old attempt made by the Famulus in *Faust,* 'to crystallise out a homunculus in a test-tube.' If any one fancies that the elements alone suffice to produce a living creature, he is making a great mistake. Even if chemistry, which is making wonderful discoveries, should ever succeed in artificially combining the same elements which are present in a living creature, and should produce

that life was diffused over the world by means of Panspermia, or germs capable of life. Although a skilful use is made of special types, this theory is forced to rely upon such fantastic subsidiary hypotheses as to seem to me altogether futile.

perfect albumen, we should still not have *living* albumen, nor a *living* protoplasm. Life would be still wanting, and it is a scientific fact that this missing residue, which we call life, cannot be accounted for at all by means of chemistry or physics. On this point I am in complete agreement with Driesch, Reinke and other Neo-Vitalists, and I regard as quite unjustifiable the statement *that Vitalism is nothing but a recourse to the unknown.*[1]

This is by no means true. The *biological fact* of life is as well known as the chemical and physical processes of life; in fact, the latter are in many respects *far less known.* But the only reasonable account that we can give of the phenomena of life is this:—There is *an internal principle,* which, in a living substance, renders the atoms, with their chemical and physical forces, capable of accomplishing something essentially higher than they can accomplish in inorganic nature. Of course, the chemical and physical forces are present—no one would deny that fact—but hitherto no one has discovered what *directs* them to the uniform aim of life, and very probably no one ever will discover it. In any case, if we desire to express the scientific opinion of the present day, we must acknowledge that *there is abundant justification for vitalism,* and the 'Autonomy of the Processes of Life,' as Driesch has formulated it, is a true *postulate* of biological science.

[1] Cf. the assertions made by Plate and von Hansemann.

Professor Plate alluded to a particularly interesting subject, viz. the alleged transitions from inorganic to organic matter in the case of *liquid crystals.* I too have read the works mentioned, especially the most important ones by Professor Lehmann, and, in reading them, it struck me at once that, in considering the points of comparison between liquid crystals and living organisms, the chief difference had been altogether overlooked. Liquid crystals grow and reproduce themselves only by taking up similar molecules from outside, whereas even the smallest and simplest of living organisms grows inwardly, and increases in size outwardly, by way of assimilation. Thus we have here the old account of life, that it tends to a purposive action *from the interior towards the exterior*, whereas in the case of liquid crystals there is only an addition of molecules or groups of molecules from outside. There is an appearance of living growth, confluence and division, but these depend upon a mere aggregation of general superficial action and of specific attraction. They are essentially *merely phenomena of disintegration.* In liquid crystals there is no assimilation of the substances taken up, so as to provide for the various needs and purposes of life such as the lowest organisms exhibit, and consequently there is *no life.*[1]

[1] Further information regarding the analogies existing between liquid crystals and living creatures may be found in Driesch, 'Bemerkungen zu Przibrams Kristallanalogien,' *Archiv für Entwicklungsmechanik*, xxiii., 1907, Part II. p. 174; also in R. Brauns' Report on Lehmann's works on

I come now to the question of *the creation of primitive forms.*

Here again my respected opponent, Professor Plate, has adopted a mistaken idea. He fancies that, in my opinion, the good God simply created for us a *primitive horse*, a *primitive ant*, a *primitive ammonite*, and so on. Have I ever said so? Can such a statement be found anywhere in my books? It occurs only in some reviews, written by prejudiced reviewers,[1] upon my work on *Biology and the Theory of Evolution*, and Professor Plate took it thence. I should like to invite my audience to examine carefully the newest edition of this book (chapter ix., section 6, p. 303, etc.), in which I made short work of all these objections. As I have already dealt with them there, I need not discuss them now, for they were merely distortions of my views. I am not so childish as to imagine the polyphyletic evolution to be so simple a matter, and such arguments prove nothing. To my mind the primitive forms of the natural orders are identical with the *primitive forms of the polyphyletic evolution*, which are recognised by many other scientists of

liquid crystals in the *Neues Jahrbuch für Mineralogie, Geologie, und Paläontologie*, 1906, ii. 2, pp. 151-153. Professor Brauns states, as the result of his investigation into Lehmann's experiments, that he was inclined to regard the changes which take place inside the molten or fluid mass as a process of disintegration, and therefore he would hesitate to lay so much stress upon the analogy with living forms. See also a Report by L. Kathariner, 'Flüssige Kristalle und Leben,' Scientific Supplement to the *Germania*, 1907, No. 24.

[1] Escherich, Forel, Haeckel. Cf. my *Biology*, chap. xii., etc.

the present day. This statement ought to suffice. They are justified from the scientific point of view, and it is absolutely unjustifiable to impute a theological tendency to them.

Professor Plate referred also to *interference on the part of the Creator.* He said that the laws of nature were laid down at the beginning of all things, and therefore there was no need for God to interfere further. This is my opinion likewise; I agree with him completely as far as the *natural* order is concerned, and this order alone was the subject of my lectures and of to-night's discussion. Professor Plate must have misunderstood me, if he imagined that our views on this point were at variance.

They certainly differ as to the origin of design in nature. Plate says that we must not assume the existence of any *immanent design,* everything must be motived from the exterior, in consequence of the struggle for existence. *This is not true,* and I might quote a number of passages, which prove its inaccuracy, from his own valuable and thorough work on Darwin's principle of selection.[1]

He abandons the interior laws of evolution whenever they are inconvenient, but the capability of reaction to external stimulus, possessed by living organisms, contains in itself these interior laws of evolution, and the principle of adaptation to purpose

[1] The passages in Plate's work to which I refer are especially pp. 14-16, 45, 51, 60-62, 142-144 (this is the most conclusive), 184-185, 188, etc., 215, etc., 224, 2nd ed., 1903.

is inseparably connected with them. I emphasised this point in my second lecture (p. 35). This *immanent directive principle* is not some vague, mystical thing, hovering at some remote height, nor is it something supernatural—we ought to have advanced beyond preconceived ideas of this kind—but it is something quite natural, it is the original constitution of the germ in question. If we require a formal principle for it, it is united with the same material substance so as to constitute *one single ens*, it is not anything supernatural—such expressions are merely phrases that ought to be set aside.

On *one point* I must acknowledge Professor Plate to be quite right. He called the world 'a vale of tears,' and I agree with him that it is so, but I think it is the fault, not of Almighty God, but chiefly of mankind! (Laughter.)

I must now refer to another topic, viz. to the conflicting principles of *Theism* and *Monism*. Here, too, *misunderstandings* have arisen which led to serious results in the case of my first opponent, as well as in that of subsequent speakers. I thought that I had expressed myself in the plainest language possible in my second lecture, but the old misunderstandings constantly recur. I wish that I could succeed in removing them once for all. Theism does not represent Almighty God as being constantly employed in keeping the machinery of the universe in motion. We regard God simply as *the origin of the natural order, as the Creator of the world,* and we assume further action on His part only where natural science or some other branch of learning constrains us to do so.[1]

This evening various speakers have raised objections to three special points.

1. *The creation of matter.* But matter cannot exist of itself, and therefore we need to have a God who created it.

2. *The existence of life.* It is only a conditional postulate which assumes the existence of God in order to account for life. If science removes the difficulty, and shows that living organisms can spontaneously proceed from inorganic matter, we shall not need to assume any interference on the part of the Creator.

3. With reference to the *intelligent soul of man*—for there is no difficulty in admitting the *possibility* of an evolution when we consider man only with reference to his body. But on the intellectual side psychology teaches us that the intelligent soul of man constitutes the *essential difference* between man and beast—and we cannot avoid this conclusion.

I may be allowed here to refer to the numerous

[1] There is no reference here to miracles, as Plate asserted in his speech. A miracle is *an exception to an already existent law of nature.* The creation of matter, the production of the first organisms, and the creation of the intelligent human soul, cannot be called miracles; to name them thus would be philosophically senseless, because in these cases the laws of nature governing these processes did not yet exist, but *still had to be imposed* by the Creator. Moreover, the laws governing the lower stages of existence are *not violated* by the laws of the higher stages, but the latter are *the complement* of the former.

arguments brought forward by Dr. Juliusburger, another of my opponents, against the immateriality and unity of the soul. I think these arguments prove nothing at all against the immaterial character of the soul, if it is rightly considered, for the soul is united with the human body so as to form *one single substance*; it is not confined within it like a prisoner in a dungeon, but with the body forms *one substance and one principle of activity.*

Moreover, all the phenomena of mental disease, mental disturbances, etc., become intelligible if we bear in mind that with regard to its activities the soul is dependent upon the functions of the organs of sense, the association paths, etc. When the nervous system suffers disturbance, the corresponding intellectual activity becomes impossible. If I were to enlarge upon this topic, I might speak for hours, and therefore I will refrain from saying any more about it.[1]

To my great joy Professor Plate stated as his own personal opinion that *behind the laws of nature there was a lawgiver* (see p. 108). Yes, ladies and gentlemen, this is a very noble statement, and I believe that our opinions on this topic approximate more closely than Professor Plate imagines. If we really regard the lawgiver as an intelligent being—and it is impossible for a lawgiver to be anything else—we have here a *recognition of a*

[1] Remarks in reply to Dr. Juliusburger's eight points will be found in the report of his speech. See p. 160, etc.

personal God, and of all the remarks made this evening, not one has given me such great pleasure as this—I could not indeed desire anything better. That we know very little by means of our natural powers regarding the nature of this God was acknowledged long ago by Christian philosophy and theology.[1]

Here again we find misunderstandings.[2]

In speaking of the rock of the Church, to which Professor Plate referred at the close of his address, he alluded to various historical matters connected with Copernicus, Galileo, etc., A word of correction is necessary. The highest ecclesiastical authority never expressed any definite condemnation of the Copernican theory. That the Congregation of the Index made a mistake at that time, every one will grant,—the Congregation is not infallible.[3]

It would perhaps have been better not to drag these subjects into our present discussion. They led us off from what was relevant to my lectures, and brought us to controversial questions.

To my regret, Professor Plate also dragged in the Reformation—a subject that I am obviously unable

[1] For the completion of this remark see p. 108 *et seq.*

[2] The reproach so often made against theism, that it pictures God anthropomorphically as a more perfect human intelligence, is explicable only by the profound ignorance regarding the Christian Theodicy which prevails among its opponents. Cf. on this subject my remarks on the speeches of Plate (p. 108), of Plötz (p. 177), of Schmidt-Jena (p. 198), and of Thesing (p. 203).

[3] In the article mentioned on p. v Dr. Burdinski interpreted this sentence as meaning that every one now grants the *Church* not to be infallible (!)

to discuss here, as it has nothing whatever to do with the matter in hand. As I stated emphatically at the beginning of my first lecture, I did not come here with the intention of touching upon religious controversy, nor did I come to wage a bitter war against popular Darwinism or Haeckelism. I came solely to throw what light I could upon the modern doctrine of evolution. I must honestly confess that I have been both surprised and pained this evening at seeing how completely my intention has been misunderstood. I am very sorry that such is the case, but I feel no personal resentment against the gentlemen whose remarks show that they have misunderstood me.

Professor Plate concluded his address with a wish for *religious unity*—I cherish similar desires in that respect, although I look for the realisation of these desires in a manner unlike that which he has in mind.[1]

Again and again this evening I have been reproached *with being inconsistent, with being fettered by dogma, with having no freedom of thought.* One speaker even went so far as to declare that I had to submit every opinion to censorship before I was allowed to express it.[2]

[1] Religious unity based on an absolute absence of creed, and on surrendering every dogma of Christianity, is an impossibility. Cf. my remarks upon Professor Plate's speech, p. 114.

[2] The speaker to whom I refer unhappily forgot to mention the right of censorship claimed by the editor of every scientific magazine, and even by the editor of every unimportant newspaper, over the intellectual productions of their contributors. It is only when a religious society exercises

But, nevertheless, ladies and gentlemen, I retain my own reasonable freedom of thought. If I submit anything to another person, and ask him to examine it and judge of its accuracy, it may well be that two heads are better than one. I have often found that the opinion of my works, which others have expressed before their publication, has prevented me from committing myself to what is false or of minor importance—and this is surely a great advantage to me, but that is only an incidental remark.

I retain my freedom of thought, provided that I submit my knowledge in *one* department to my own knowledge in *another* and higher department.[1]

When, therefore, Professor Plate declared at the end of his speech that *I could not be a genuine scientist or a genuine scholar*, he was expressing his own private opinion, which I at least do not share. (Laughter.)

. . . .

My audience will pardon me for dealing more shortly with my next opponent. Dr. Bölsche expressed his own views on the subject of monism, and he was quite free to do so; but if he by

any control over the publications of its members that the cry of 'Intolerable subjugation of the intellect' is suddenly raised.

[1] If any one is convinced that one truth cannot contradict another truth, he will regard it as a matter of course for a scientist, who is at the same time a theologian, to try to reconcile his scientific with his theological knowledge. A scientist who has no religious belief is certainly not justified in asserting that he alone aims at truth, simply because he is an unbeliever. I was glad to perceive from Professor Dahl's closing remarks (p. 134) that not all modern scientists are so narrow-minded.

any chance imagines logic to be a strong point in monism, as he understands it, I regard that as extremely doubtful, and am inclined to think that those have more logic on their side who assume the existence of a Creator, or, as Professor Plate says, of a lawgiver, who originally laid down the laws of nature. This seems to me far more logical.

The alleged gradually increasing differences between man and beast, and the intellectual evolution of man from the animal kingdom,—these are subjects which could only be treated adequately in a special lecture on Comparative Psychology. I have published several works on them, to which I must refer my audience.[1]

My opinions therefore coincide with those of Dr. Bölsche only when he praises logic, and I wish very much that scientists in particular were well endowed with it. (Laughter.)

. . . .

Professor Dahl, whose remarks appealed to me very much, believes that in the eternity of matter there is no opposition to *the laws of thought*, but only to the *imagination*. It is undoubtedly true

[1] A long reply to Bölsche would be superfluous, as all that he said about the psychology of animals, and the excellence of the animal soul in comparison with the human soul, was from the point of view of 'Popular Psychology.' I may refer my readers to my own works, *Instinkt und Intelligenz im Tierreich* (Instinct and Intelligence in the Animal Kingdoms) 3rd ed., Freiburg im Breisgau, 1905, and also *Vergleichende Studien über das Seelenleben der Ameisen und der höheren Tiere* (Comparative Studie, of the Mental Activity of Ants and of the Higher Animals), 2nd ed., Freiburg im Breisgau, 1900. Cf. also my remarks on Bölsche's speech, p. 119.

that we have a difficulty in imagining the creation, but to imagine the eternity of matter is, of course, impossible. But, as I said before, the impossibility of accepting the theory that matter is eternal does not depend upon our imagination at all, but it resides in the *philosophical principle* that only an infinitely perfect being can of itself exist for all eternity. Matter is not infinitely perfect, therefore it cannot be eternal. We have to deal here with philosophical contraries.

I have one more point to mention. Professor Dahl referred to *the intellectual development of a child*, in order to support his theory that the human soul might develop out of the animal soul by a process of natural evolution. His argument is very plausible, but it must be borne in mind that, in the case of this ontogenetic development of a child, *there is present always one and the same intelligent soul*, which gradually reveals its faculties as the powers of the intellect are evolved, and this evolution is essentially dependent upon that of the nervous system. If a young ape began, at the age of six or seven, to express itself intelligibly, we should have some evidence of the possibility of the evolution of the human from the animal soul—but the development of a human child does not seem to me to furnish this evidence.

. . . .

I come next to Dr. Friedenthal, and I wish to begin by saying that I was very glad to hear him state definitely this evening, that his chief aim was simply to demonstrate *the chemical and physiological resemblances existing between different kinds of blood.* We are therefore of one mind, and the popular idea that he interpreted kinship of blood to imply a common origin or descent was based on a misconception. His explanation was one that I gladly accept. He said also that his investigations were concerned with man only as being the highest type of mammal. This is true in my opinion also ; with respect to his body, man represents the highest form of mammal, and therefore he may zoologically be classed as standing next to the order of apes.[1]

Dr. Friedenthal said : '*We are not only descended from apes, but we are ourselves genuine apes.*' If by these words he means : 'With regard to our bodily organisation we stand in immediate proximity to the Primates,' he is only stating a well-known fact, with which even Linnæus was familiar, whom I do not wish to contradict. But I do contradict any one who maintains that the actual evidence of the descent of man from beasts is so well established as to force us to accept it. This is the real difficulty. I do not challenge the assertion that it is possible for man, with regard to his body, to be descended from beasts—let me emphasise the word *possible*—and in so doing I leave the theological question

[1] I do not say that he is to be classed in the same order as apes, as Haeckel, Friedenthal, and others propose to do.

quite on one side, for we are not now concerned with it.

. . . .

I turn now to Professor von Hansemann, and again have to begin by pointing out a misunderstanding. He fancies that I have declared the problem of the evolution of animals not to be merely a zoological question. I never have said such a thing; on the contrary, I have always maintained that it is a purely zoological question, in itself quite independent of every theory as to the cosmic position of man, whether theistic or monistic.[1]

With reference to the relation in which *Religion* and *Science* stand to one another, Professor von Hansemann taunted me with not being able to anticipate the decision of the Church. That is true. As a theologian I am forbidden to do so, but as a scientist I may go on quietly, with no fear of interference. I am not tied down to any fixed course, for one truth cannot stand in the way of another. Here again we have a confusion of ideas.

As to the occurrence of appendicitis among savages I shall perhaps be in a position to give some information to the Professor later on. It would certainly be very interesting if we could prove from statistics that these inflammations are, for the most part, a result of hypercivilisation.[2]

[1] Cf on this subject pp. 6, 19, 22.

[2] In my remarks on Professor von Hansemann's speech, p. 153, I have given my reasons for not regarding the vermiform appendix as a rudimentary organ.

. . . .

We come next to the objections raised by Count von Hoensbroech, but in my opinion *not one of them was relevant to the subject of to-night's discussion.* I will therefore refrain from making any further comment upon them; they concerned questions of religion, and were out of order. (Loud applause—a few hisses.)

. . . .

I have already dealt with the points mentioned by Dr. Juliusburger, who is both a psychologist and an alienist (p. 223 and p. 160, etc.). His arguments tend to disprove the existence of an intelligent soul. I should have to speak for a long time, if I were to expound clearly the philosophical conception of an intelligent soul, and I cannot begin to discuss such a topic to-night. I should like, however, to make one remark upon *the Theory of Identity*, upon which Dr. Juliusburger based his arguments. The theory has been satisfactorily disposed of by Privy Councillor Stumpf in his lecture on 'Body and Soul,' which was the inaugural address given at the International Psychological Congress at Munich, on August 4th, 1896. Cf. also my criticism of Forel's theory of identity in *Instinct and Intelligence*, 3rd ed., chap. xii.

The monistic avowal, which Dr. Juliusburger made here to-night, of the identity of all existing things, seems to me to correspond more with certain emotions than with logic. I cannot blame any one for saying that he accepts this theory, but personally I cannot reconcile it with clear, philosophical thought, to adopt monism, *i.e.* the essential identification of God and the world, for the acceptance of this theory involves us in innumerable contradictions. As soon as we assume God to be essentially identical with the world, He shares in all the imperfections of the world, and this contradicts our idea of God as the infinitely perfect being.

. . . .

I can only express my gratitude to Herr Itelson for his kindly anxiety about the rock of Christianity, and I can assure him that I, for my part, have no intention of crumbling away from the rock. (Laughter.) On the whole, his words showed his goodwill to me, as he called my coming forward here a ‘consoling appearance.’

. . . .

In answer to Dr. Plötz, I may single out what he said about the *Pithecanthropus* for special comment. He referred to difficulties which he thought I had not taken sufficiently into consideration. The time allotted to me for my lecture on the subject was too short for me to mention every possible difficulty, but more explicit information may be found in my book on *Biology and the Theory of Evolution.* 3rd ed., p. 474. As to Macnamara's cranial curves, to which Dr. Plötz alluded, many other considerations have to be taken into account besides the size and capacity of the cranium. Men like Kollmann,

Kramberger, and even Schwalbe himself, who are authorities on anthropology, are of my opinion to this extent, that they think the Pithecanthropus is not a direct ancestor of man. In the case of the Neandertal cranium, which I pronounced to have belonged to a human being of some lower race, Dr. Plötz thinks I did not lay sufficient stress upon the presence of superciliary ridges and the absence of chin. I mentioned these points in my book; there was not time to discuss everything during my lecture. I may, however, state clearly once for all that, among the numerous crania examined by Kramberger, there is no uniformity on these points—viz. the superciliary ridges and the absence of chin—but there are successive transitions leading up to modern man.[1]

The so-called *Homo primigenius* therefore proves

[1] In several respects the opinions of the various anthropologists are widely divergent. Kramberger (whom I quoted in my third lecture, p. 75) thinks that genuine superciliary ridges and instances of absence of chin occur sporadically as individual variations among men of the present day, the blacks in Australia, etc., whereas Schwalbe does not admit this to be the case ('Studien zur Vorgeschichte des Menschen,' *Zeitschrift für Morphologie und Anthropologie*, extra number, 1906). It is true that the receding chin seems to be the most important racial characteristic of man of the early diluvial age. Cf. C. Toldt, 'Zur Frage der Kinnbildung,' *Korrespondenzblatt der Deutschen Gesellschaft für Anthropologie, Ethnologie, und Urgeschichte*, xxxvii., No. 2, February 1906, pp. 9-17. As Dr. Hugo Obermaier proved in his work on the earliest remains of the human body studied from the point of view of anatomy and anthropology (Vienna, 1905): '*With absolute certainty we can only say that man of the quaternary period differed in no essential respect from man of the present day. In no way did he go beyond the limits of variation of the normal human body.*' In body, as well as in mind, he was already a genuine *homo sapiens*. Cf. also Obermaier, *Der diluviale Mensch nach seiner intellektuellen (kulturellen) Seite*, p. 11 *et seq.*

to be only a man belonging to an earlier race than the *Homo sapiens* of the present day.

. . . .

I was particularly glad that Dr. Schmidt-Jena spoke so calmly and impartially. I had been forced by circumstances to express myself rather strongly in my lectures, not against Haeckel personally, but against the confused ideas which, in my opinion, were due to Haeckel's influence; and consequently I was afraid that Dr. Schmidt-Jena might speak with some animus. Although I had done my best to avoid using any offensive expressions, it was quite possible that some few of my remarks might have aroused feelings of annoyance, and therefore I was extremely glad that my opponent used such moderate and impartial language.

I will now turn to the objections raised by him.

In the first place, he declared that I had misunderstood Haeckel on some points, that Haeckel had not been confusing Darwinism with the doctrine of evolution for the last forty years, and that he had, moreover, brought forward his pedigrees only as hypotheses. I think we must make a distinction here; there are two personalities also in Haeckel. (Laughter.)

I ask your indulgence if I apply to Haeckel also what has been said of myself, viz. that I possess a twofold nature, and am at once a theologian and a scientist. Yes, indeed, there are in Haeckel two personalities: he is at once a scientifically accurate student of nature and a bold advocate of Darwinism, dealing freely in generalisations. He has both these personalities; and not unfrequently he himself mistakes one for the other. In his *General Morphology*, and still more in his later work on *Systematic Phylogeny*, Haeckel certainly has taken pains to draw a tolerably clear and precise distinction between the theory of evolution and Darwinism. In these works he speaks of the theory of evolution as a construction formed of hypotheses. But he expresses himself in quite another way very often when he is using popular language and addressing the general public. I happen to have by me the oration which he delivered at Cambridge in 1898, before an assembly of zoologists, and subsequently published for the benefit of wider circles. It is entitled: 'Our Present Knowledge of the Origin of Man' (*Ueber unsere gegenwärtige Kenntnis vom Ursprung des Menschen*), Bonn, 1899. On p. 22 is a passage which I propose to read aloud to you, as bearing upon the charge brought against me of having misunderstood Haeckel's pedigrees, whereas Haeckel had no intention of bringing them forward as dogmatically correct, but merely as modest hypotheses. Haeckel asserts the contrary with reference to his Pedigree of the Primates, which he gives here on p. 35, and which I criticised in my third lecture.

On p. 22 we read:—'The general outlines of the

pedigree of the Primates, from the earliest eocene Lemures up to man, fall within the tertiary period and are perfectly plain. There is no essential link still missing. The phyletic completeness of the race of Primates from the oldest Lemures to man is an historical fact.' [1]

These words stand here, and Haeckel himself had the last sentence printed in italics. This is the language that he uses in addressing the general public. I must acknowledge that in many places he has modified his statements, especially in the last few years—but, nevertheless, what I said before still holds good.

With regard to the *biogenetic principle*, Dr. Schmidt-Jena accused me of having reproduced it in an incomplete form, but I was obliged to be brief, and to content myself with remarking shortly that the repetition of the evolution of the race was influenced by adaptation. Dr. Schmidt-Jena will find the matter discussed at greater length on p. 458 of the last edition of my *Biology and the Theory of Evolution*. I have there distinguished

[1] That this quotation is perfectly accurate can be proved by any one who takes the trouble to refer to the passage. Moreover, it is obvious that in it Haeckel lays down the pedigree of the Primates '*up to man*' *as an historical fact.* Nevertheless, a reporter of my closing address accused me, in the *Berliner Morgenzeitung* of February 20th, 1907, of having falsified the quotation. He declared that, in quoting this passage, I passed off what Haeckel had really said about the pedigree of apes as being his teaching about the pedigree of man. Either the reporter was not aware that he, as man, belonged to the Primates (according to Haeckel), or for other motives, which will not bear criticism, he chose simply to impute falsification to me.

palingenesis and cænogenesis, and have shown that they cannot be reconciled in that so-called fundamental law. In my lecture I assumed that the biogenetic principle was generally known, and therefore I did not discuss it in detail, knowing that here in Berlin I was addressing a highly educated audience.

Dr. Schmidt-Jena and I are diametrically opposed on the subjects of Theism and Monism. He regards Monism as *a deduction from the doctrine of evolution*, but I showed in my first lecture that *the doctrine of evolution as such, being a scientific hypothesis and theory*, is not in any way concerned with any theory as to the cosmic position of man. I might with equal right declare theism to be a deduction from it; in fact, I should have better justification for so doing, for, in order to account for the origin of primitive forms, whether their number was great or small, if we view the matter from a scientific and philosophical standpoint, we are forced to assume the action of *some principle other than spontaneous generation*, if this does not admit of scientific demonstration. I cannot avoid the conclusion that theism, even when one tries to support it by scientific arguments and to compare it with monism, proves to have a firmer foundation. *It is untrue* to say that the doctrine of evolution, regarded as a scientific hypothesis and theory, necessarily leads to monism and pantheism. Such is *not* the case.

Dr. Schmidt-Jena went on to say that he thought

the rock of Christianity was in a rather precarious condition, and he felt bound to infer that it was in a state of retrogression or disintegration. Well, I think I need not discuss this point, for some mean one thing and some another in speaking of this rock.[1]

When I referred to it at the conclusion of my third lecture, I meant what is common to all Christians, and in some sense also to the Jews, viz. the theistic view of man's position in the universe, as this view has existed in its historical form for two thousand years. What meaning has been assigned to it by others in the course of to-night's discussion is quite indifferent to me.

I must, however, refute one charge that Dr. Schmidt-Jena brought against me when he asserted that I acknowledged myself to be an adherent of *deism* rather than of *theism.*[2]

Theism accepts the existence of a personal God, who created the world and laid down the laws of nature, and allows the world to develop independently by refraining from any arbitrary interference with the laws which He has imposed. This is quite correct, but, in adopting Theism, we must not overlook the fact that it represents this God as present in all creatures, and as participating in all their actions. This is not Deism, nor is it Pantheism. It sounds like Pantheism, but why? Because Pantheism borrowed it from the old theistic idea of

[1] Cf., for instance, the remarks made by Dr. Bölsche, p. 117.

[2] For a full explanation of these terms, see my remarks on Dr. Schmidt's speech, p. 198.

God. It is nothing more than what St. Paul expressed in the words: 'In Him we live and move and have our being.'

. . . .

As to my last opponent, Dr. Thesing, I was heartily glad to find that I had at least one friend and protector on some few points present to-night. (Laughter.) Therefore I do not intend to discuss the difficulties which he raised, as I have dealt with them adequately, as I think, in my answers to previous opponents.

I conclude therefore by assuring you once more that I came to Berlin to offer you, in these lectures, *such information as I could give in a calm, impartial way, and not to carry on a war of words.* (Loud and prolonged applause. The proceedings ended after midnight.)

SUPPLEMENT

THE long discussion of February 18th, 1907, was over. It was represented in some newspapers as having been a great intellectual conflict. What was its result? Did I and my opponents arrive at any *understanding* regarding our various opinions? No, *such was not the result* of the discussion, nor was such a result possible, considering the circumstances. I believe, however, that not only I, but also my opponents, originally intended in the course of the discussion simply to state our various views plainly and dispassionately. Circumstances unfortunately frustrated their good intention. On this topic I may refer to the view expressed by an impartial witness on my opponents' side, Professor Dr. H. Potonié, who retired on the day of the discussion.[1]

In the *Naturwissenschaftliche Wochenschrift* for 1907, No. 10, p. 157, he writes: 'Just as *Father Wasmann stated his opinion in a very brilliant fashion*, especially in the three lectures that preceded the evening discussion, so, I think, are we too in a position *to indicate our present point of view*, and to allude only to the matters on which we differ.' This was said by an editor of the *Naturwissenschaft-*

[1] Cf. p. 87.

liche Wochenschrift, who acknowledges that '*he personally is in direct opposition* to Father Wasmann.' I am quite willing to agree with Professor Potonié in believing that this evening discussion could not have sufficed to bring about a mutual understanding and a harmonising of our points of difference; it would require years to effect such a result, as he remarks in his report. Professor Dahl, too, was perfectly right when he said, at the end of his address at the evening debate, that in his opinion it would be more expedient to attempt an explanation of our divergent views in writing. Professor Dahl expressed his desire that such an attempt should be made, and others of my opponents, especially Professor Plate, the chief speaker, would have done well to disregard all that did not strictly form part of the subject of my lectures, and not to impute 'theological intentions' to me, when I was absolutely devoid of them. Why was this not done? I believe it was chiefly because the discussion was *not limited to specialists*, speaking before specialists, as I had originally intended, but it was extended to the general public.[1]

[1] I proposed in the first instance to connect the discussion with a private meeting of the German Entomological Society, which took place on February 16th. Dr. Walter Horn, the President of the Society, who was also on the Committee that organised my lectures, had invited to this meeting the chief scientists, and especially the zoologists, of the University, of the High School of Agriculture, and of the Royal Natural History Museum. Another member of the Committee, however, rejected my proposal, and insisted upon a public discussion. The meeting was very well attended, and its interesting proceedings closed with cheers for the Jesuit Father Wasmann. Cf. the report in the *Germania* of March 2nd, 1907.

If the original plan had been carried out, probably only prominent scientists would have taken part in the discussion, whereas at the public meeting speeches were made which were not at all to the point, such, for instance, as those of Count von Hoensbroech and of Itelson the author, and this was almost inevitable. A tempting opportunity presented itself for turning the scientific discussion of Father Wasmann's lectures into *a hostile demonstration against them in the name of free scientific research, or, in other words, into a modern religious argument.* Without being a prophet, any one might have foretold that such an argument could lead to no mutual understanding.

Who can claim to have been victorious in this religious argument which took place on February 18th, 1907, in the great hall of the Zoological Gardens, in this argument to which, contrary to my wishes, the scientific discussion was reduced ?

The answer is important. In all previous religious arguments both parties have claimed the victory, and we must therefore see what impartial experts said on the subject. Whatever the result might have been, it was a foregone conclusion that the *Vossische Zeitung*, the *Berliner Morgenzeitung*, and other similar journals would claim a 'brilliant victory won by free scientific research over ecclesiastical trammels.'[1]

[1] Burdinski did the same thing in his already-mentioned article (see Preface, p. v), *Der Kampf um die Weltanschauung in Berlin* (The

It is worthy of notice that *the chief proof* alleged of Wasmann's defeat is the fact that the eleven speakers were all opposed to him, not one took his side, and of the general public no one came forward to support him. I have explained this fact adequately in dealing with the circumstances that led up to the evening discussion. I was under the necessity, either of retiring altogether from the discussion, or of accepting the conditions imposed by the majority. The list of speakers was given in the programme which the President read out, and it was therefore obvious that my numerous supporters among the audience could express their views only by their applause at the end of my final speech. I have already dealt with the distortions of truth, with which I have been charged by the *Berliner Morgenzeitung* and a few other papers that echo its sentiments. It is a sufficiently significant fact that certain persons have had to have recourse to such means as this, in order to weaken the force of my answer.

The impression made by the evening discussion upon impartial listeners seems to have been, that the violent attacks of my opponents *did not succeed* in destroying the effect produced by my three lectures. The *Freisinnige Zeitung* actually affirms this, and says that my antagonists' lances were unable to

Struggle in Berlin regarding the Cosmic Position of Man) p. 40. He used, in fact, the same words as the *Vossische Zeitung*, but there is nothing to be gained by a further consideration of his article.

lift Father Wasmann out of his saddle. Although the *Frankfurter Zeitung* was decidedly opposed to my appearance at Berlin, yet it acknowledged that my opponents had not succeeded in overthrowing *me*, and the counter-arguments suggested as more likely to silence me were unfortunately too late. Another non-Catholic paper, the *Deutsche Tageszeitung*, went further than any Catholic journal in its criticism of several of my opponents, saying (No. 84, of February 19th): —'The result of the meeting will have been a disappointment to many. If we except Professor Plate, whose profound knowledge is united with great facility of expression and an earnest striving after accuracy,[1] those who took part in the discussion appeared like pygmies beside Father Wasmann, and the mild ridicule with which he finally answered them would have been bitter satire in the mouth of another reporter. There is no doubt that the audience would have dispersed soon after Professor Plate's address, if the majority of the vast number present had not been anxious to hear Father Wasmann's reply to his antagonists, and had not regarded it as a duty to express, at the close of the

[1] Another Protestant critic, Dr. M. Senff, in the *Harzer Kurier* of April 27th, 1907, does not share this opinion, but considers Plate's whole line of argument to be prejudiced, and not free from inadmissible interpolations (cf. p. 251). The description of the speeches quoted above cannot be accepted without modification, for some of them, *e.g.* those of Professor Dahl and Dr. Juliusburger, surpassed that of Professor Plate in accuracy of matter and of form. The comparison quoted above seems therefore not altogether to the point.

proceedings, their gratitude to the learned Father for all his exertions and explanations, by means of hearty applause.'

Another non-Catholic paper, the evangelical *Christliche Welt* (No. 12, March 21st, 1907), reproaches me with having confused religion and science in my lectures. Ernst Teichmann, one of the reporters, declared that I understood by 'science' only ecclesiastical dogmas based on scholasticism, but this statement was perfectly unjustifiable,[1] as any one would acknowledge, who had either heard or read my Berlin lectures. The following admission of the same reporter seems to me particularly worthy of notice, in comparison with his other statement.

'Whoever reads the outlines of Wasmann's views on the theory of descent (as given previously in detail by Teichmann), and seriously considers them, must be impressed by the fact that, in the opinion of their advocate, they are one and all based upon the foundation of strictly scientific methods. We may think what we like of their value, but we cannot deny that Wasmann has given an absolutely logical account of his scientific views, in accordance with his own understanding of the matter, and with the means at his disposal. It was impossible therefore to single out any one of his statements in order to confront it with another, and to reveal any inconsistency between them. Such a line of action

[1] Cf. also the opinion of another Protestant reporter, Dr. M. Senff, who is quoted on p. 253.

will never lead to success . . . and the criticism on Wasmann's views, passed by the scientists assembled at Berlin, was futile.'

A Catholic theologian in Berlin, in some remarks on my Berlin lectures which he contributed to the *Allgemeine Rundschau* (Munich, March 16th, 1907), expressed the opinion that, both in my lectures and in my closing speech at the evening discussion, I had touched too lightly on philosophy and theology.[1]

On the other hand, I was accused by a reporter in the *Hochland* of April 1st, 1907, of having dwelt too much upon philosophy and theology. The truth lies perhaps midway between the two extremes. The author of the criticism in the *Hochland*, a non-Catholic, concludes his article with the following words:

'The disgraceful fact remains that Wasmann, an insignificant priest, in consequence of his training, and not of his intellectual abilities, speaking as a philosopher, routed our collective scientists, and in the course of the discussion displayed the greatest tact in combating that scientific arrogance, which deals with truths that are limited to an existence of twenty-five years, as Ibsen-Stockmann tells us, whereas the Church, in her exalted wisdom, is fully conscious that no earthly truth, of any kind whatever, can be contrary to a divine truth.'

The Berlin *Tägliche Rundschau* (No. 85, Feb.

[1] Dr. Leo Heidemann replied to this criticism in the *Rundschau* of March 30th.

20th, 1907), actually attempted to accuse President Waldeyer, who took the chair at the discussion, of being prejudiced in my favour. Professor Waldeyer felt himself obliged to protest against this groundless accusation in a subsequent number of the same journal (No. 105, March 3rd.)

My intention in alluding to the matter is only to show to what means a certain section of the press had recourse, in order to represent the proceedings during the evening discussion from their point of view.

The most conclusive condemnation of the action of a large proportion of the so-called liberal papers was expressed in a letter that appeared on March 28th in No. 13 of the *Israelit, Zentralorgan für das orthodoxe Judentum,* which letter appeared in an abbreviated form also in the *Germania* (No. 83, April 12th.).

As a result of the war of words which began before my first lecture in Berlin on Feb. 12th, and was continued for months in the liberal papers, with reference especially to the evening discussion, the *Israelit* concludes, not altogether unreasonably, that '*Liberalism and Intolerance in religious matters are identically the same thing.*' [1]

[1] The number of newspaper articles dealing with my Berlin lectures and the evening discussion, or connected with them, already exceeds five hundred. I cannot waste more time upon them. Even in the *Kladderadatsch* and in the *Jugend* there were references to me. Among the most harmless results of my lectures, we hear of a 'newly discovered kind of sea creature,' nearly related to seals, which was described in the April number (No. 13) of the *Berliner Illustrierte Zeitung.* It is worth notice that

I consider it quite unnecessary to enter upon any detailed discussion of the attacks upon me which the *Vossische Zeitung* continued to publish for three months after my lectures. I will allude only to one article, which appeared on April 26th, and I do so merely because it completely bears out the opinion expressed in the *Israelit*.

The *Vossische Zeitung*, speaking as the organ of scientific men, allows itself to say:

'The point at issue (in the dispute with Father Wasmann) *is not whether the Jesuits are right in any one department of science, but whether the multitude regards them as right in their efforts against progress.* We need not repeat what is at stake. Everything can be summed up in the word *Counter-Reformation.*'

In reply, the *Germania* of April 27th remarked with considerable bitterness:

'It comes to this: On scientific questions a Jesuit can never be right, just because he is a Jesuit. Even if occasionally a Jesuit should really be right on a scientific point, he must not be allowed to stand his ground, for his efforts are contrary to progress; and if that statement does not suffice, we are told that the liberty won by the Reformation is at stake. In the eyes of the multitude he is thus completely defeated, and we, the disciples of true science, are freed from the ungrateful task of refuting

very few of the numerous criticisms published by my opponents expressed any sympathy with Haeckel and his Monism.

him on scientific grounds according to the laws of logic. O Freedom of thought! O impartiality! O true, O German Science! O city renowned for Intelligence! can such things be said in thee without rousing thee to indignation?'

I feel bound to plead the cause of true German science and its representatives in Berlin in reply to this harsh condemnation. Among the chief scientists in Berlin, I have made the acquaintance of men of profound learning, and equally great tolerance in religious matters. I need only mention the names of Wilhelm Waldeyer and Oskar Hertwig. These men, and others like them, showed a really scientific and impartial interest in my lectures, and the unworthy persecution which the *Vossische Zeitung* and similar papers aroused did not proceed from the kings of science, but from their inferiors, who presumed to express the general views of German and especially of Berlin scientists.

. . . .

Let us now return to the evening discussion. *What took place* at it, and *what was its result?* Before expressing my own opinion, I should like to quote the views of Dr. M. Senff, a Protestant critic, who contributed to the *Harzer Kurier* of April 27th and 28th an article on this subject, filling almost ten columns, entitled 'The Jesuit Father Wasmann, for or against?' From this article I have taken the following extracts. Dr. Senff criticises particularly Professor Plate's speech, describing him as the most important speaker extensively and intensively on the opposite side.

'Professor Plate's polemic,' writes Dr. Senff, 'has its weak points. It is unfair to say that Father Wasmann rejects the theory of descent in reference to man because of his ecclesiastical prejudices. Scientists who are not Jesuits are daily coming to the same opinion; and excellent reasons for this rejection, and perhaps the best of all, are derived from quite another source. Any one who is free from prejudice is bound to acknowledge that Father Wasmann has a right to put forward his arguments against the theory of evolution. He has given sound arguments in abundance, which have nothing at all to do with Roman orthodoxy, since they are defended even by Protestant scholars. What is fair to one, should be accepted by the other. In my opinion Plate made *an unwarrantable assertion* when he said that, in the case of Wasmann, the scientist was always subordinate to the theologian in his arguments. Professor Plate seems to me to have assumed the existence in Father Wasmann of a strife between science and theology, in order the more easily to attack him. It cannot be denied that the Theory of Evolution, when extended to man, leads in the case of a Jesuit to a serious conflict with the Church, but if Wasmann claims to be judged as *a scientist* and not as a *Jesuit*, scientific etiquette requires us to comply with his desire, as long as he really adheres to science and brings in no extraneous Jesuitical arguments. We should have a right to complain of such, *if he had substituted dogmatic opinions for scientific reasoning*, but, as far as I can judge, he has not done this, and has, on the contrary, brought forward sufficient evidence of a kind *that would not suggest the idea of a religious conflict to any unprejudiced person*, supposing—as is daily the case—it were a Protestant scientist who adduced it in support of his views. Therefore, I repeat, what is fair for one, should be permissible to the other.'

'All that Professor Plate adds in this connection *seems to me equally prejudiced, and not free from an inadmissible misinterpretation, if not an actual setting aside of Father Wasmann's arguments.*'

'It was not surprising that Plate's polemical speech was applauded by a public that was incapable of forming a judgment, being already prejudiced in his favour—such a public always applauds what takes its fancy. It is all the worse in my opinion—*there is a suggestion of something not quite straightforward about the proceedings, and this does not please me.*'

Dr. Senff passes on next to the subject of 'bad metaphysics.' He analyses Professor Plate's views and finds them very vague and contradictory, for at one moment Plate acknowledges that behind the laws of nature there must be *a lawgiver*, and at the next moment he assumes that matter and the laws governing it are uncreated and *everlasting.* 'I think,' says Dr. Senff, 'that considerable obscurity exists here, and that Plate's rationalistic views on natural science may be very attractive to a modern audience, that desires to be entertained, but they will prove on more exact analysis to be unsound and incoherent.' 'Professor Plate is a believer—this he has frankly acknowledged in spite of all his science—but it is difficult to see why he stops half-way; it would be more consistent if he went a step further. Let us have either all or nothing! This obvious tendency to stop half-way is nothing but the outcome of superficial thought, of modern prejudice with regard to natural science, and of inadequate acquaintance with the requirements of logic, philosophy, and metaphysics. It is the patching up of a personal opinion after the fashion of modern thinkers. In this instance, too, we feel that our great philosophers have once more lived in vain.'

Dr. Senff next passes on to Plate's tirade against *the creation of living creatures.* He first examines the reasons for assuming the existence of a living principle, and arrives at this conclusion: 'We cannot avoid the recognition of some wonderful and inexplicable formal principle of a transcendental and metaphysical nature.' 'The reproach of *being prejudiced* ought not to be brought against those who—like Father Wasmann—acknowledge the recognition of a formal principle to be inevitable, but rather against those who—like Professor Plate —yield to the tendency of modern thought to assign a natural and mechanical origin to everything, and allow their superficial kind of philosophy to inspire them with the hope of avoiding such a recognition.'

This statement applies more especially to the first origin of a living creature out of inorganic matter. Plate's arguments in favour of spontaneous generation are rejected by Dr. Senff as having no more value than Haeckel's. 'I am sorry to be obliged to follow the Jesuit, rather than the professor of zoology, on the point under discussion.'

Dr. Senff regards the alleged transitions between what is inorganic and what possesses life as merely external analogies, that cannot prove any essential equality between them. 'Professor Plate is as little able, as any one else, to bring forward a single case, in which it is certain that the line dividing the inorganic from the living has given place to continuity uniting them.

'What, then, is the use of juggling with analogies and parallels, that prove nothing, but only dazzle an audience which is eager to applaud, and neither disposed to nor capable of any exact examination ? The subject is too serious and too important to be dealt with in this way, which is, after all, not true science, but *very fine dogmatism* on the part of students of natural science.'

The critic goes on to consider the relation in which *scientific research* stands to *metaphysics.* 'Father

Wasmann freely and frankly stated that he ascribes its full importance to metaphysics, a confession which must at all costs be impressed upon the conscience of modern humanity.' 'This brings us to the decisive issue, to the most important point of difference between "believers" and pure scientists. Father Wasmann claims the right to survey, from the standpoint of a philosopher and metaphysician, the isolated facts which have been brought to light by experimental, empirical, and exact research, and he regards it as not only permissible, but as actually prescribed, in order to attain to a more profound understanding of the problems of life, to have recourse to all possible means, and to tolerate the imposition of no limits. Professor Plate, on the other hand, condemns all this in a jealous way as an unwarrantable encroachment, and finds fault with it as an untrustworthy blend of inadequate materials, as the importation of ideas that serve only to obscure, disturb, and defile the subject, as a spurious kind of science based upon prejudices.'

Plate's assertion that in drawing philosophical conclusions from scientific facts, I was influenced *by ecclesiastical prejudices*, is absolutely denied by Dr. Senff, who says: '*Has Professor Plate clearly established the charge which he brings against Father Wasmann of having in any case started from ecclesiastical prejudices?* Wasmann's opponents represented him as having taken this course, and we know that many hounds are death to the hare. I am vexed that Professor Plate and others waged an easy war upon *the Jesuit*, when they ought to have aimed their attack at *the student of scientific research.* In my opinion Wasmann *did not start with any ecclesiastical prejudices, but as a scientist engaged in research work*, he has arrived at results which are not necessarily opposed to his religious belief.[1]

'*He is quite within his rights in joyfully proclaiming this fact*, nay, it is *his duty* to do so, if he desires the welfare of his fellow-creatures. I may here state my standpoint more definitely. The demand for a clear separation between scientific research and philosophy is justified only to the point, where exact observation of isolated instances ceases, and leads to some general result. To connect and appreciate such results (which are in themselves worthless either in isolation or in co-ordination) is the task of philosophy, and I do not see how any progress could be made, if a scientist might not be a philosopher. Separation has therefore its *limits*, and in my opinion, in spite of the many interesting speeches made by his opponents, *it has not been proved that Father Wasmann has outstepped these limits.*

'**I should prefer somewhat less ecclesiastical**

[1] A similar opinion is expressed by another Protestant reporter, Dr. Beth, in the *Neue Preussische Zeitung* of May 9th. He emphatically states that 'Wasmann's fundamental theory and general attitude with regard to the doctrine of evolution need not be ascribed to any subservience on his part to Church or to dogma, but have in his case the same empirical foundation as in the case of a number of modern scientists, who cannot be accused of rejecting Darwinism in its more special sense, *in order to avoid incurring episcopal censure, or acting in a way contrary to dogmatic tradition.*'

indignation and somewhat more scientific truth —*even although it may be inconvenient. If scientific truth prevailed, the awkward situation would not arise that, in a Protestant country, a sense of honour compels a third person to come to the assistance of a Jesuit.*'

Dr. Senff proceeds to refute Plate's assertion, that, in assuming the work of God in the production of the first living being and in the creation of the soul of man, I have violated any natural law. He says: 'The higher obedience to law, with its naturally higher differentiation, always includes in itself the laws of the lower stages, without breaking or violating them or even setting them aside.'

This, then, is the view adopted by up-to-date philosophers, men of eminently clear and sane judgment, who are assuredly neither mystics nor obscurantists.

I limit myself to these quotations from the criticism passed *by a Protestant* upon the evening discussion. He is *an impartial witness*, one who is not a Catholic priest and a Jesuit, and so one who cannot be charged *with lack of freedom of thought and of ability to follow things to their logical results*, through having his way barred by fear of ecclesiastical censure.

. . . .

In conclusion, I may sum up shortly, under three headings, the proceedings at the evening discussion and their results:

1. My eleven opponents *did not, collectively, succeed in encountering and refuting me* on the ground of scientific facts, and of the philosophical deductions from them.

2. Some of my opponents strayed from the subject of my lectures, and turned, what professed to be a scientific discussion of them, into *an attack upon the Catholic Church,* doing this in the name of free science.

3. If the supporters of free research were able to combat my statements only in this way, they have provided me with the best possible proof that these scientific and philosophical opinions regarding the theory of evolution, which I, as both a Christian and a scientist, have put forward, *do not clash with the principles of really free research.*

SUPPLEMENTARY NOTE UPON pp. 171, 218, 231.

In the *Naturwissenschaftliche Wochenschrift,* No. 27, July 7th, 1907, is a treatise, entitled 'A Contribution to the Question, What is Life?' by Dr. Dahl, who was the third speaker at the evening discussion. In this treatise he lays stress upon the fact that there is nothing mystical in the expression 'vital force,' but that it is only another name for the properties peculiar to living creatures. Some of his statements regarding *psychical life* are particularly worthy of notice. From the standpoint of his own experience, Dr. Dahl speaks very decidedly against the monistic theory of identity. The processes of movement, that go on in the brain, remain always something totally different from the phenomena of consciousness. If we regard their actual connection as constituting 'an identity, as the monists do, we at once leave the *terra firma* of experience, and find ourselves in the region of mysticism. We are unable to adduce a single fact, based on experience, in support of their identity.' . . . 'However much we may struggle against dualism, we cannot avoid its acceptance, if we abide strictly by what experience teaches, as it behoves students of natural science to do.'

. . . .

I have just received a work entitled *Ultramontane Weltanschauung und moderne Lebenskunde, Orthodoxie und Monismus. Die Anschauungen des Jesuitenpaters Erich Wasmann und die gegen ihn in Berlin gehaltenen Reden, herausgegeben von Prof. Dr. Plate.* Berlin. Mit 12 Textfiguren. Jena, 1907. Gustav Fischer. (Ultramontane Views of the Position of Man in the Universe and the Modern

Theory of Life, Orthodoxy and Monism. The Opinions of Father Erich Wasmann, S.J., and the Speeches made in Opposition to him in Berlin, edited by Dr. Plate. Berlin. With 12 Illustrations. Jena, 1907.)

This work gives an abstract of my three Berlin lectures; the accompanying illustrations have almost all been borrowed from my book, *Modern Biology and the Theory of Evolution.* The speeches of my opponents are given, some in an expanded form, and then my closing address, which is much abbreviated, and, lastly, some remarks by the editor. I will only say a few words about this rejoinder. My readers will be able to judge for themselves who has truth on his side.

Professor Plate's introductory remarks about the evening discussion (p. 3, etc.) have been sufficiently dealt with in my statement on p. 85, etc., of this work. On p. 10 he asserts that I refused to publish my lectures jointly with my opponents, 'because I said that I had been badly treated by them, and insisted upon special conditions of publication.' The terms proposed to me in writing, by Professor Plate, would have deprived me of what I was plainly entitled to claim, viz. the right to reply to my opponents' speeches (which had lasted three hours) more fully than it was possible for me to do in my closing address, delivered at midnight on February 18th. It will be obvious to any one that I could not accept such conditions.

I see, on reference to Plate's work, that, in the present publication, the contents of my opponents' speeches have been reproduced exactly and correctly, with the exception of those made by Count von Hoensbroech and Mr. Itelson, which have been omitted as irrelevant. I will not complain that Professor Plate, in the new version of his speech, has considerably expanded and modified it. His statement, quoted verbatim on p. 109, to the effect that it was only logical to assume the existence of a lawgiver behind the laws of nature, is, on p. 70 of Plate's own publication, weakened down by the addition of the words, 'whom I imagine as a very high intelligent principle in the pantheistic sense.'

The critical remarks in my present work will serve to refute the opinions of my opponents, even in their present form.

I need not discuss the 'final considerations' which Professor Plate added to his book, because they are made, not from the point of view of objective science, but from that of a partisan expressing his hostility to the Church. To them, even more than to Professor Plate's speech, applies the criticism of Dr. M. Senff, a Protestant, quoted on p. 249, etc., which closes with the words:

'I should prefer somewhat less ecclesiastical indignation, and somewhat more scientific truth.'

INDEX

Adaptation, 43.
Adaptive variation, 11-13.
Agnosticism, 109.
Amical selection, 47, 132.
Ancestry of man—
  Branco's views on, 78.
  Haeckel's views on, 80, 125, 191.
  Plötz's views on, 181, 232.
  Schwalbe's views on, 74, 78.
Anergates, 12.
Ants—
  parasitic, 12.
  red robber, 47.
  slave-maker, 12.
  tertiary, 10.
  variation among, 13.
  See also Inquilines.
Ape, skeleton of, compared with human, 56, 68, 181.
Apes, fossil, and lemures, 72, 78, 191.
Apes and men—
  alleged links between, 72, 73, 180.
  blood-relationship between, 66, 136, 229.
  Friedenthal's views on, 138-143, 229.
  Plötz's views on, 181, 232.
  theories as to connection between, 66, etc.
  See also Archiprimas, Archipithecus, Pithecanthropus, Primates, Prothylobates.
Appendix vermiformis, 65, 153, 230.
Archipithecus, 80.
Archiprimas, 80.
Atheism, 25, 200.
Atomistic philosophy, 147.

Biogenetic principle, 44, 58, 70.
—— Schmidt-Jena on, 191, 236.
Blood, investigations regarding, 66, 134.
Blood-relationship between men and apes, 66; Friedenthal on, 136, 229.
Brachyoptera, 12.
Branchial arches and clefts, 60, 61.
Branco, Professor, 78, 101.

Cænogenesis, 59, 193, 237.
Cerebral formation, human, 56, 182.
—— functions, localisation of, 163-167.
Chromosomes, 44.
Church not hostile to science, 93, 210.
—— unity of the, Plate on, 112, 225.
  See also Rock, Freedom of research.
Clavigeridæ, 13.
Columbus, 93, 94.
Copernicus, 3, 82, 113, 145, 158, 224.
Cosmozoic theory, 208, 215.
Cranial capacity of man, 181-186.
—— formation of men and apes, 56, 74, 233.

Creation—
biblical account of the, 17, 54.
Plate on, 102, 113.
Plötz on, 177.
—— conception of, 26, 28, 33, 97, 178, 212.
—— of matter, 96, 97, 178, 213, 222.
—— theory of, not equivalent to theory of permanence, 4, 18, 103.
See also Matter, Organisms, Spontaneous generation.
Creator, true conception of a personal, 26, 105, 109, 178, 198, 204, 206, 213, 221.
Creator, false conception of—
Dahl's, 126.
Haeckel's, 27, 204.
Plate's, 105, 220.
Plötz's, 177.
Schmidt-Jena's, 197.
Thesing's, 203.
Crystals, liquid, 100, 218.

DARWINISM, 37-48.
—— Dahl on, 132.
—— Schmidt-Jena on, 187, etc., 232.
See also Theory of selection.
Deism, 199, 238.
Descent of man. *See* Man.
—— theory of. *See* Evolution.
Design in nature, 26, 35, 107, 131, 150, 220.
—— von Hansemann on, 150, 152.
—— Plate on, 107, 220.
Dinarda, 11.
Doryloxenus, transformation of, 11.
Double nature, Haeckel's, 187, 191, 234.
—— Wasmann's (according to Plate), 94, 115, 211, 248.
Dual consciousness, 169.
Dualism, 24, 172, 258.
Dubois, 73.
Dysteleologies, 151, 155.

ENTELECHIES, 30. *See* Vitalism.
Evolution—
causes of, 42-44, 219.
cosmic, 4.
laws of, 29, 35, 43.
monophyletic, 15, 16, 104, 203.
polyphyletic, 14, 16, 104, 203, 219.
principles of. *See* causes of.
proofs of, 8-15.
scientific theory of, 1, etc.
in relation to the Bible, 16-20.
in relation to Christianity and philosophy, 23, etc., 103, 144, 210, 237, 250.
in relation to Darwinism, 37, etc., 132, etc., 183-189.
of man and the higher animals, 51, 58, 126.
See also Man, descent of; Monism, Theism, Views of life (Weltanschauung).

FAITH, foundation of, 110.
Fleischmann, 15.
Formal principles, 30.
Formicoxenus, 13.
Freedom of research, 54, 55, 77, 225, 250, 254, 257.
—— Bölsche on, 121.
—— von Hansemann on, 145.
—— von Hoensbroech on, 158.
—— Plate on, 94, 116.
—— Senff on, 248.

GALILEO, 93, 224.
God, conception of—
Haeckel's, 27, 204.
Linnaeus', 37.
Monistic, 25, 109, 123.
Plate's, 108, 109, 223.
Plötz's, 178.
theistic, 21, 24, 109, 131, 198, 213, 221.
Thesing's, 203, etc.

HERTWIG, OSKAR, 37, 59, 127, 194, 216, 249.
Homo primigenius, 75-80, 233. *See* Neandertal man.
—— *sapiens*, 77.
—— *sapiens fossilis*, 77.
—— *sapiens primigenius*, 77.
Hypotheses, indispensability of, 24, 116. *See also* Freedom of research.

IDEALS, human—
Bolsche on, 120.
Friedenthal on, 137.
Identity, monistic theory of, 172, 231, 258.
Inflammation, a beneficial process, 155.
Inquilines among ants, 11, 12, 46, 47.
Insane, treatment of the, 173.
Instinct and intelligence, 156.
Intellectual activity—
definition of, 31.
Bölsche on, 120.
Dahl on, 128.
Friedenthal on, 140.
von Hansemann on, 156.
Juliusburger on, 163.
Plötz on, 184.

KRAMBERGER, 75, 233.
Krapina cranium, 75, 119, 120, 186.

LAMARCK, 43, 187, 188.
Lawgiver in nature, 108, 223, 259.
Lehmann, 100, 218.
Life, nature of. *See* Vitalism.
—— origin of. *See* Organisms *and* Spontaneous generation.
—— principle of, 10, 30, 147, 202, 218, 252, 258.
Localisation of cerebral functions, 163-167.
Logic, monistic, 118.
—— universal (Bölsche), 121, 227.
Lomechusa, 47.

MAN, descent of, 49-80, 126-128, 236.
—— —— Bölsche on, 119, 227.
—— —— Dahl on, 126, 228.
—— —— Friedenthal on, 140, 229.
—— —— Haeckel on, 79, 190, 235.
—— —— Plötz on, 176-185, 232.
—— —— Schmidt-Jena on, 190, 236.
——, soul of, 31-33, 49, 53.
—— —— application of evolution to, 49-57, 128.
—— —— palæontological proofs of, 72.
—— —— psychological aspects of, 53.
—— —— scientific aspects of, 54, 55.
—— —— theological aspects of, 54.
—— —— zoological proofs of, 55.
—— —— Bölsche on, 118, 227.
—— —— Dahl on, 128, 129, 228, 258.
—— —— Friedenthal on, 140.
—— —— von Hansemann on, 156.
—— —— Juliusburger on, 160-176, 222, 231. *See also* Descent of man, *Homo primigenius*, Neandertal man, Intellectual activity.
Matter, creation of, 29, 33, 212, 222, 251. *See also* Creation.
—— eternity of, 34.
—— —— Dahl on, 123, 228.
—— —— Plate on, 96, 212.
—— —— Plötz on, 178.
—— —— Senff on, 251.
—— —— Thesing on, 206.
Melancholia, 168.
Mental disturbances, 167-170, 223.
Metaphysics, in relation to scientific research, 251, 252.
Mimeticon, 12, 14.

Mimicry among inquilines, 12, 46.
Miracles, 105, 111, 222.
—— Plate on, 111, etc.
—— Schmidt-Jena on, 196, etc.
Missing link, 72, etc.
—— Plötz on, 180, 232.
Monism—
  meanings of, 24, etc.
  metaphysical, 24.
  postulates of, 25.
  views of, Bölsche's, 118, etc., 226.
  views of, Juliusburger's, 171-176.
  views of, Plate's, 108, 221.
  views of, Schmidt-Jena's, 195, 237.
Monophyletic evolution, 15, 16, 104.
Morphology of man and ape, 55, 64, 70, 139, 143, 180, etc. *See also* Ape *and* Cranium.
Mutation, theory of, 9, 39.

NATURALISM, 199, 252.
Natural selection, 38, 106, 132, 152, 187, 201, 220.
Nature, laws of—
  intelligent author of, 34, 109, 223.
  in relation to miracles, 105, 111, 198, 222.
  Bölsche on, 121.
  Plate on, 109, 223.
  Schmidt-Jena on, 197.
  Senff on, 256.
Neandertal man, 74, 181-186, 233.

ONTOGENESIS, 191, 192.
Organisms, origin of, by creation, 29, 99, 104, 218, 222, 252.
—— Dahl on, 129.
—— Plate on, 97, etc., 214, etc., 218.
—— Thesing on, 207.
Organisms—
  See also Cosmozoic theory *and* Spontaneous generation.
Organs, rudimentary, 63-66, 152, etc.
Origin of life. *See* Organisms *and* Spontaneous generation.

PALINGENESIS, 59, 193, 237.
Pantheism, 199, 238.
Pathological arguments against design in nature, 150-156.
Paussidæ, 12, 13.
Pedigrees—
  in the theory of evolution, 6, 15, 18, 103, 219.
  Haeckel's, 79, 190, 235.
  of Primates. *See* Primates.
  See also Man, descent of.
Permanence, theory of, 4, 18, 82.
—— Plate on, 103.
Philosophy in relation to evolution, 23, 96, 211, 245, 246. *See also* Metaphysics.
Pithecanthropus alalus, 80.
—— *erectus*, 56, 72, 73, 74, 80, 180, 232.
Polyphyletic evolution, 16, etc., 104, 203, 219.
Potonié on the Evening Discussion, 240.
Press opinions on the Evening Discussion, 240, etc.
Primates, Haeckel's, 79, 191, 235.
Primitive types, 101, 219.
Processus vermiformis. *See* Appendix.
Prothylobates, 80.
Pselaphidæ, 12.
Psychology, 53, etc.
—— monistic, 171, 231, 258.
—— Bölsche's, 119, 120.
—— Dahl's, 128, 228 258.

Psychology, Friedenthal's, 140, etc.
—— von Hansemann's, 156.
—— Juliusburger's, 160-176, 231. *See also* Intellectual activity, Theory of identity, Soul of man, Theory of transformation.
Purpose in nature. *See* Design.
Pygostenus, transformation into inquiline of termites, 11.

REACTION, power of, in protoplasm, 35, 43, 220. *See* Vitalism.
Rock of Christianity, 81.
—— Bölsche on, 122.
—— Itelson on, 159.
—— Plate on, 112, etc.
—— Schmidt-Jena on, 196.
Rudimentary organs, 63-66, 152, etc.

SELECTION—
sexual, 133, etc.
theory of. *See* Natural selection.
Senff on the Evening Discussion, 244, 249-256.
Skeletons of apes and men, 56, 68, 181.
Soul, animal, 31, 53.
Bölsche on, 119, 227.
Dahl on, 128.
Friedenthal on, 141.
von Hansemann on, 156, 157.
—— human. *See* Man, soul of.
—— simple nature of, 160-176.
Species—
conception of, 17, etc.
natural, 18, 102, 219.
Plate's supernatural, 101.
systematic, 4, etc.
Spontaneous generation, 25, 29.
—— Dahl on, 129.
—— Plate on, 97, etc., 214, etc., 253.
—— Thesing on, 207.
Symphilic type of inquilines, 12, 46.
Sympolemon, 12.
Synechthry, 12.

TEICHMANN ON THE EVENING DISCUSSION, 245.
Termites, 9-13.
Termitomyia, 13, 63.
Termitoxenus, 13, 62.
Thaumatoxena, 13.
Theism—
doctrine of personal Creator, 21, etc., 26-37, 47, 81, etc., 104, etc., 198.
distinguished from deism and pantheism, 198, 199, 238.
Dahl on, 124.
Plate on, 108-110, 223.
Plötz on, 170, etc.
Schmidt-Jena on, 196, etc., 238.
Thesing on, 203, etc.
Transformation, monistic theory of, 171, etc.
Trilobitideus, 12.

VERMIFORM APPENDIX, 65, 153, 230.
Views of the universe (Weltanschauung), 23, etc.
—— monistic, 23, etc., 108, 109.
—— theistic and atheistic, 23, etc., 82.
—— Darwinian, 37, etc.
—— Haeckel's, 197.
—— Christian, 23-37, 47, 81.
—— —— Bölsche on, 122.
—— —— von Hansemann on, 144, etc.
—— —— Itelson on, 159.
—— —— Juliusburger on, 174, 175.
—— —— Plate on, 108, 109, 210 etc., 237.
—— —— Plötz on, 176, etc.

Views of the Universe (Christian)—
—— —— Schmidt-Jena on, 195, etc., 237.
—— —— Thesing on, 203, etc. *See also* Theism, Monism, Deism, Pantheism, Naturalism.
Vitalism, 30, 202.
—— Dahl on, 258.
—— von Hansemann on, 148.
—— Plate on, 106, 217, 252.
Vitalism, Thesing on, 202.

WHALES, evolution of, 62.

ZOOLOGY, competent to deal with evolution of animals, 4, 144, etc., 230.
—— *not* competent to deal with evolution of man, 49, etc., 144, etc.